Colli

Pira

David Pickering

First published in 2006 by
Collins, an imprint of
HarperCollins Publishers
77-85 Fulham Palace Road
London, W6 8JB

www.collins.co.uk

Reprint 10 9 8 7 6 5 4 3 2 1 0

A catalogue record for this book is available from the British Library

Created by: SP Creative Design
Editor: Heather Thomas
Designer: Rolando Ugolini

ISBN 13 - 978-000-723007-5
ISBN 10 - 0-00-723007-9

Printed and bound by Amadeus S.p.A., Italy

CONTENTS

INTRODUCTION

Images of pirates are deeply entrenched in popular culture. The most universal image is that of a dashing, cutlass-wielding adventurer who defies unworthy enemies against a backdrop of sunlit seas and desert islands. In the popular imagination, the instantly recognizable features of pirate life include such things as skull-and-crossbones flags, eyepatches, buried treasure, pieces-of-eight and walking the plank, but the reality was far from romantic. Most pirates were greedy, treacherous robbers who preyed indiscriminately on defenceless merchant vessels. Violence was a way of life, and some took pleasure torturing or killing prisoners.

There have been pirates for as long as men have sailed the oceans. Their victims have ranged from the sea traders of ancient Greece and Rome to Spanish treasure galleons and East Indiamen laden with the riches of the Indian sub-continent. Few pirates made their fortune: most faced lives of extreme hardship and danger that ended prematurely in disease, drowning, murder or even execution.

Nonetheless, pirates have always appealed to the imagination. A few individuals, such as Sir Henry Morgan, Blackbeard, Captain Kidd and the women pirates Anne Bonny and Mary Read, have entered legend and the real facts of their lives have been forgotten. This book aims to reveal the truth behind the popular image.

TYPES OF PIRATES

The word 'pirate' describes any person who seeks plunder on the high seas, illegally attacking shipping of any nationality. The term includes the following sub-categories.

Privateers

A 'privateer' (or 'privateersman') was a pirate who (like Captain Kidd) held a 'letter of marque' which licensed him to attack the ships of a national enemy during wartime.

Buccaneers

A 'buccaneer' was a pirate of the seventeenth century who attacked Spanish shipping and possessions in the Caribbean. Famous English buccaneers included Sir Henry Morgan.

Corsairs

A 'corsair' was a privateer or pirate who attacked shipping in the Mediterranean. Most feared of all were the Barbary corsairs of north Africa.

PART ONE

Chronology

Pirates have roamed the seas of the world since ancient times. Generations of peaceful seafarers have learned to dread the appearance of pirates, and piracy itself has long been one of the most colourful threads running through maritime history.

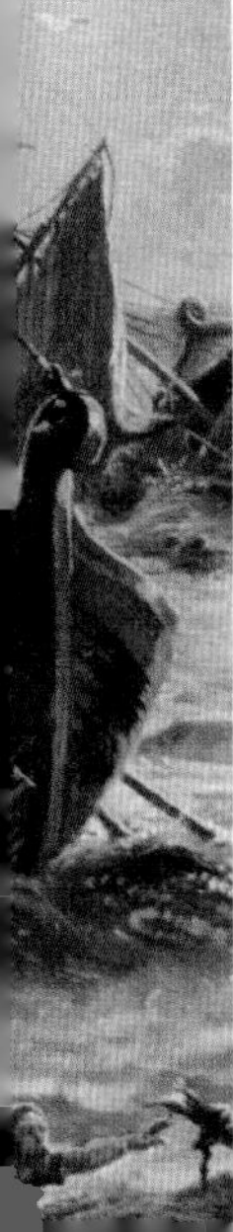

Viking longboats
These Scandinavian sea raiders were also intrepid sailors, navigating the Atlantic Ocean and venturing as far afield as the Mediterranean, Greenland and probably even North America in their sleek, square-sailed boats .

PIRACY THROUGH THE AGES

The earliest records of piracy go back thousands of years. Seafarers in the Mediterranean were falling victim to pirates even before the emergence of the ancient Egyptian civilization. Pirates continued to plague the Mycenean, Minoan, Phoenician, Greek and Carthaginian civilizations until they were suppressed by the Romans in the first century BC.

Piracy in the Mediterranean re-emerged during the period of the Byzantine Empire. The waters of northern Europe, meanwhile, were terrorized by Saxon and then Viking warrior-pirates. Maritime trade became busier in the medieval period, and with it came an increase in piracy. Islamic corsairs sailing out of the Barbary coast roamed the Mediterranean, while privateers were employed by various European monarchs.

SWASHBUCKLERS

The treasure ships carrying the wealth of the New World to Spain inspired an explosion in piracy in the so-called Spanish Main in the sixteenth century. This was the era of the swashbuckling privateers, buccaneers and English sea dogs, whose adventures became the stuff of legend.

The 'Golden Age' of piracy, however, opened late in the seventeenth century, when pirates such as Blackbeard terrorized the Caribbean and Indian Oceans under the skull-and-crossbones flag, earning fearsome reputations for cruelty and greed. Although most of them were dead within 20 years, they are still remembered in the popular image of the historical pirate, complete with cutlass, eyepatch and chest of buried treasure. The daring of the French corsairs added to the romantic image of pirates, as did the bravado of American privateers like John Paul Jones during the American Revolution and the War of 1812.

PIRATES OF THE FAR EAST

Few pirates were as vicious as those who swarmed the South China Sea and the Indonesian archipelago in the eighteenth and nineteenth centuries. Pirate junks dominated sea lanes, and the power of the pirate chiefs rivalled that of the Chinese emperors themselves.

Piracy today

Piracy was curtailed in the mid-nineteenth century, but has never disappeared completely. Even today reports come in of acts of piracy off Africa and in the South China Sea.

THE SEA PEOPLES

The so-called 'Sea Peoples' were a loosely-linked confederation of piratical tribes who dominated the eastern Mediterranean around 1200 BC. Although little is known in detail about them beyond mentions in the hieroglyphics of Egyptian temples, it appears that they were nomadic and came originally from various parts of Asia Minor and elsewhere in the Mediterranean region. They probably included the feared Lukkan sea raiders, who had plagued coastal trade routes in the Mediterranean since the fourteenth century BC, making the trading ships of ancient Egypt a particular target.

LORDS OF THE MEDITERRANEAN

These early pirates used highly manoeuvrable and fast oared galleys and made trade between early civilizations in the Mediterranean a perilous business. No trading vessel sailing in the Mediterranean or in the Persian Gulf was safe from interference.

The Sea Peoples appear to have forged alliances with Egypt's enemies from time to time and the Egyptians fought pitched battles with them on at least one occasion (off the Nile Delta in 1186 BC). At the height of their power they commanded the whole of the eastern Mediterranean, attacking shipping and conducting coastal raids. Major cities and even entire regions fell under their control. After their defeat by the Egyptians most of the Sea Peoples appear to have settled in Palestine and turned to farming, possibly becoming the ancestors of the Phoenicians.

The influence of the Sea Peoples is thought to have had a significant effect in slowing the development of ancient civilization in the Mediterranean region.

This engraving shows an early sailing vessel, complete with sails and oarsmen.

PIRATES OF ANCIENT GREECE

Pirates were a threat to shipping in the eastern Mediterranean throughout the period of the Ancient Greeks. Several Greek city-states, including those of the fourth-century BC Aetolian League, adopted piracy to protect their trade against their neighbours, allowing pirates to establish lairs in their territory. Others built up fleets to suppress the pirates who attacked their ships.

One Ancient Greek legend relates how pirates once kidnapped Dionysus, the Greek god of wine. When Dionysus transformed himself into a lion the pirates jumped into the sea, where they were themselves turned into dolphins.

The Aegean Sea, with its numerous islands and inlets, proved ideal for piracy, offering havens where pirates could shelter. Particularly notorious were the Dorian Greek pirates who overwhelmed the Minoans in the tenth century BC and made Crete a pirate base for the next 800 years. Another major pirate lair was Lemnos, from which pirate fleets raided Athens itself.

TREASURES OF THE ANCIENT WORLD

The richest prizes included Phoenician traders carrying silver and precious metals, although pirates also attacked humble fishing boats and conducted coastal raids. Any captives were sold as slaves.

Like pirates of all ages, those of the ancient world used whatever vessels they could obtain, though they preferred fast, shallow-bottomed galleys powered by oars which were easy to manoeuvre and could cross shallow waters where larger vessels could not follow.

Piracy declined with the establishment of large navies by the Greek city-states and the campaigns of Alexander the Great. The pirates of Lemnos were dispersed in the sixth century BC, while those of the islands of Kithnos, Mikonos and the Sporades were suppressed in the following century. A similar fate befell the pirates of Crete in the second century BC.

ROME AND THE PIRATES

Having been suppressed by Alexander the Great and others, the pirates of the Mediterranean enjoyed a revival in fortune with the decline of Macedonia and the outbreak of war between Rome and Carthage in the second century BC. At different times large fleets of pirate ships were employed as mercenaries by one side or the other. Regions such as Dalmatia and Illyria embraced piracy as a means of enriching themselves at the expense of Roman and other shipping. Cilicia (in what is now southern Turkey) and various islands in the western Mediterranean, such as Corfu and Cephalonia, became notorious pirate lairs.

The pirates typically used fast oar-powered bireme galleys

Julius Caesar
As a young man, Julius Caesar was captured by Cicilian pirates while sailing to Rhodes. After five weeks in captivity he was released for a ransom. Caesar later attacked the pirates who had taken him and had them crucified.

The pirates' fast galleys often had eyes painted on the prow for luck. There were plenty of Roman merchant ships for them to attack and cargoes of grain, wine and olive oil to be seized. Pirates also found a ready market for prisoners and kidnap victims in the Roman slave markets of Delos and elsewhere.

CONQUEST OF THE CILICIAN PIRATES

The Roman army conducted several campaigns against pirates in the Mediterranean, including those of Illyria, Istria and Dalmatia, and ultimately drove them out. Pompey the Great finally broke the power of the Cilician pirates with a huge fleet in 67 BC, and Cilicia itself was seized by the Roman army. By the time of Emperor Augustus, the Mediterranean was largely clear of pirates and it was not until the decline of the Roman Empire early in the fifth century that piracy emerged as a significant factor again.

THE VIKINGS

The Vikings were fierce sea raiders from Scandinavia who terrorized the coastline of northern Europe for some 300 years from the late eighth century AD. The appearance of Viking longboats caused panic among local populations, who had good reason to fear these sword and axe-wielding warrior-pirates with their long beards and battle-loving reputations.

The Vikings were skilled navigators and boatbuilders and their square-sailed longboats took them all round northern Europe and beyond, reaching Russia and the Mediterranean and probably landing as far away as the coastline of distant north America. Their preferred method of battle was to swarm ashore from a beached longboat, catching their enemies by surprise, or else to come alongside another vessel and board it. Such was their success that few merchant vessels dared to venture on the seas for fear of being intercepted.

FEARLESS RAIDERS

Calling on Odin and the rest of the Norse gods, the Vikings prided themselves on knowing no fear, and their raiding parties often penetrated far inland in search of plunder. Any captives taken could expect little mercy, and priests and peasants alike were slaughtered or carried home as slaves. Villages and towns were commonly torched. Some communities only saved themselves by paying large ransoms.

Eventually, around the early twelfth century, the Norsemen mostly gave up raiding and settled in the lands to which they had sailed, turning to maritime trading and other more peaceful pursuits.

The term 'Viking' actually means 'going on an overseas raid'.

BYZANTINE PIRATES

The Byzantine Empire emerged from the ruins of the Roman Empire in the fourth century AD to dominate the eastern Mediterranean for 1,000 years. Although piracy was quite rare, the empire declined and came to depend increasingly upon the help of mercenary pirates to fend off seaborne enemies, particularly after the collapse of its own navy.

After Constantinople was sacked in 1204, pirates operated throughout the Aegean Sea. Occasionally they forged new alliances with the Byzantines against the growing power of Venice and Genoa. In the late thirteenth century, the pirates (most of whom were Italian, Greek or Balkan by birth) acted as Byzantium's navy, attacking Italian shipping in the Aegean.

Eclipsed by the corsairs

In the fourteenth century, the influence of the Byzantine pirates was eventually contained by competition from the Ottoman Turks, who employed their own fleets of Islamic pirates – the corsairs of the Barbary Coast.

MEDIEVAL PIRATES

Maritime trade grew steadily, providing opportunities for pirates. The Hanseatic League, formed by German ports on the Baltic in 1241, tried to keep piracy down but after its decline shipping in the area became vulnerable. An English version of the Hanseatic League, called the Cinque Ports, started out with similar aims but restricted itself to protecting English shipping and attacking ships of other nations.

RIGHTS OF REPRISAL

The outbreak of the Hundred Years' War between England and France provided opportunities for English pirates, who preyed on French and Spanish shipping. In response, the French and Spanish granted their own pirates the right of Reprisal, authorizing them to seize English shipping in compensation.

Government protection

Eustace the Monk was protected by the English crown when plundering French shipping, but when he attacked English ships he had to flee to France and was employed by the French.

BARBARY CORSAIRS

The Barbary coast of North Africa became home to one of the most feared of all pirate confederations in the late fifteenth century when Islamic corsairs serving the Ottoman Empire set up bases in Tunis, Algiers, Tripoli and other local ports.

Over the next 150 years the Barbary corsairs attacked non-Muslim targets throughout the Mediterranean, seizing shipping and conducting raids on the coast as well as protecting Islamic trade routes. Although their motivation was partly material profit, it was also partly loyalty to their Muslim rulers, who came to rely upon their share of pirate income.

The corsairs of the Barbary coast sailed in long, sleek, swift galleys and smaller, even nimbler galiots.

Christian corsairs

Curiously enough, the ranks of the Barbary corsairs included many renegade Christians, who adopted the Islamic faith in order to share in the riches to be had from taking part in the corsairs' piratical activities.

The Barbary corsairs were feared by Christians throughout the Mediterranean for their brutality.

Brutal pirates

The corsairs had a reputation for brutality among Europeans, who considered them as the worst of pirates, although they were seen as legal privateers and heroes in the Islamic world.

Any prisoners whom the corsairs took were likely to be pressed into service as oarsmen on their galleys or sold as slaves. When major battles were fought between the Ottoman navy and their Christian enemies, corsair vessels usually made up a significant part of the Muslim fleets.

Celebrated corsair leaders included the Barbarossa Brothers, Murat Rais and Uluj Ali. The network of corsair bases was managed by a council called the Taife Raise, which answered in turn to the Ottoman Empire itself. The Ottomans for their part provided Turkish troops to defend the corsair ports, especially from attack by Christian Spain.

The Barbary corsairs were less of a threat after the mid-seventeenth century but remained active for another 150 years or more until they were finally suppressed by combined European and American fleets early in the nineteenth century.

HUGUENOT PIRATES

In the early sixteenth century the Huguenots (French Protestants) were engaged in a religious war with Catholic Spain in the Americas, and French Huguenot captains realized they could make a fortune by attacking Spanish treasure galleons sailing back from the Caribbean. After the French established a Huguenot pirate base at Fort Caroline in Florida in 1564, the Spanish ordered nobleman Pedro de Menéndez de Avilles to drive out the pirates. He seized the fort and massacred the pirates who fell into his hands, destroying the power of the Huguenots in the New World.

Raids on the Spanish

Led by such captains as Jean Florin (or '*Fleury*') and François de Clerk (nicknamed *Jambe de Bois* or 'Pegleg'), French pirates attacked French shipping and coastal settlements on both sides of the Atlantic. They conducted numerous raids on the coast of Hispaniola and even captured the cities of Havana and Santiago in Cuba. The Spanish were forced to strengthen their defences throughout the New World.

THE CORSAIRS OF MALTA

The piracy of the Barbary corsairs threatened the welfare of the famous Knights of Malta. In response the Knights established their own force of Christian corsairs to protect their interests, assembling crews from Malta, Corsica and France. With the Knights as captains of their vessels, these corsairs believed they had a mission to rid the Mediterranean of heathens, as well as making money for themselves. The greatest victory against their Muslim enemies was in 1565 when they drove off an Ottoman fleet sent to capture Malta.

THE CORSAIR FLEET

The fleet of 30 carracks financed by the Knights provided employment for up to a third of the island of Malta's population. The carracks themselves shared much in common with the Barbary galleys, but tended to carry more sail and a larger number of guns. Like the galleys of the Barbary corsairs, they were oar-powered, the oarsmen being largely crew members from captured ships.

The corsairs of Malta gradually ceased operations from the 1680s in the wake of treaties between various European and Barbary rulers.

THE PRIVATEERS

Some pirates sailed as 'private men-of-war' (privateers) with official backing. In 1243, Henry III of England granted Adam Robernolt and William le Sauvage permission to attack his enemies, on condition that they divided any booty taken equally with the crown.

As well as documents permitting captains to harass enemy shipping in time of war, some licences allowed merchants who had lost ships or cargoes to make good their losses by attacking ships of the nation concerned.

Some privateers, such as Sir Walter Raleigh and Sir Francis Drake, were rewarded by the monarch for the wealth they secured for themselves and their nation. England was not the only country to embrace the principle: French privateers included René Duguay-Trouin, who enjoyed the support of King Louis XIV.

Legitimate or illegal?

The line between privateering and piracy was often blurred and it is not surprising that what one country considered legitimate privateering another considered outright piracy.

THE SPANISH MAIN PIRATES

The Spanish Main was the term originally applied to the seas off the northern coast of South America but later included all the Caribbean and surrounding land masses. The first Spanish colony in the Caribbean was on the island of Hispaniola (now divided into Haiti and the Dominican Republic). The *conquistadors* probed deep into the area around the Caribbean basin and set about plundering the treasures of the defeated Aztec and Inca nations. The waters of the Spanish Main were regularly traversed by Spanish treasure ships and consequently by the various pirate vessels that sought to seize their valuable cargoes.

THE PIRATE SEA

With so much rich trade to attack and no shortage of remote anchorages in which to hide, the area acquired a reputation for piracy. Despite Spain's dominance in the region until the early seventeenth century, ports such as Havana, Cartagena, Vera Cruz and Panama were not safe from attack by pirates. The Spanish tried to deter them by reinforcing settlements and transporting treasure in huge convoys of as many as 100 ships, but still their galleons continued to provide rich pickings for English, French and other seafarers.

THE SEA ROVERS

In the 1570s, a new generation of English pirates became the scourge of the Spanish Main. Often claiming to be privateers sailing on behalf of the English crown, they filled their ships' holds with booty from Spanish treasure ships and coastal settlements. The treasure was divided between the pirates and the crown, although Queen Elizabeth I was officially unaware of their activities and any support she provided usually had to be given in secret.

NATIONAL HEROES

For 20 years the sea rovers disrupted the transport of gold and silver from the Spanish colonies in the New World and threatened the financial stability of Spain. Their raids were thus in the English national interest, even though many of the pirates were probably as much inspired by the chance of enriching themselves as they were by notions of patriotism.

Some of the so-called 'sea rovers', such as Sir Francis Drake, Sir John Hawkins, Sir Martin Frobisher and Thomas Cavendish, became national heroes back in England and were honoured accordingly by their queen, much to the fury of Spanish ambassadors.

THE BUCCANEERS

The term 'buccaneer' came from the French *boucan* (barbecue), which referred originally to the way in which French settlers living on Hispaniola (now Haiti and the Dominican Republic) used to smoke the meat from the pigs and cattle they hunted. These tough backwoodsmen, who were the first to be known as buccaneers, were bitter enemies of the Spanish, who had driven them from their homes.

In the 1620s, these French buccaneers began to attack Spanish ships sailing close to the island of Hispaniola, using canoes and other small boats to steal up quietly to larger ships and then board them before the crew realized what was happening. They quickly earned a reputation for ruthlessness and cruelty to those who resisted them.

THE BROTHERHOOD OF THE COAST

The early buccaneers were soon joined by many others, among them runaway slaves, convicts, adventurers and seamen of various nationalities, especially the English. Despite Spanish efforts to destroy this growing menace, some large buccaneer communities were established on the island of

Freebooters and filibusters

French buccaneers were also called 'freebooters' or 'filibusters' (after the swift, but small *flibotes* or 'fly boats' they used).

Tortuga, at Port Royal in Jamaica and elsewhere in the Caribbean, providing the bases from which the so-called 'Brotherhood of the Coast' terrorized shipping of all nations throughout the Spanish Main over the next 50 years.

Raids on the Spanish

From the late 1650s onwards, the buccaneers felt confident enough to conduct raids on Spanish towns and cities, such as Santiago in Cuba and Porto Bello in Panama. The names of their leaders, who included Sir Christopher Myngs, Sir Henry Morgan and the notorious Frenchman François L'Olonnais, became feared throughout the Caribbean and beyond.

The end of an era

The eventual declaration of peace between England and Spain in 1670 effectively meant the end of the buccaneering tradition, but the term buccaneer continued to be used more loosely to describe any pirate who sailed the high seas.

PIRATES OF THE CARIBBEAN

Between 1690 and 1730, cutthroats and brigands roamed the Caribbean and the eastern coastline of America, including such infamous pirates as Edward Low, Blackbeard, 'Black Bart' Roberts, Charles Vane, Stede Bonnet, Howell Davis and the women pirates Anne Bonny and Mary Read.

THE GOLDEN AGE OF PIRACY

Some ports became notorious pirate lairs, while the pirates earned evil reputations with their wild and often murderous behaviour. Their careers were usually short, commonly ending in death by violence, execution or disease, and few were able to retire on the wealth they had obtained. Piracy was much reduced in the 1720s, largely due to pirate hunter Woodes Rogers, who became governor of the pirate island of New Providence in the Bahamas in 1720.

It has been estimated that there were around 700 pirates active in the Caribbean in the so-called 'Golden Age'. Most were English by birth, but they also included Spaniards, French, Dutch, Americans, West Indians, Scots and Welsh as well as runaway black slaves of African descent.

INDIAN OCEAN PIRATES

With the end of the buccaneering era, some pirates decided to sail further afield in search of plunder and ventured round the Cape of Good Hope and into the Indian Ocean, lured by the fabled wealth carried by the treasure fleets of Arabian princes and the Indian Moghul Empire. There was also the chance of seizing merchantmen belonging to the British, French and Dutch East India Companies that sailed eastwards with gold and silver to purchase the riches of Asia, with which they then returned westwards.

The Homeward-Bound Indiaman. Those ships carrying luxury goods from India back to Europe became obvious targets for many pirates.

A PIRATE PARADISE

The tropical island of Madagascar, off east Africa, proved an ideal pirate base, situated close to the trade routes between Europe, India and China. It soon became a notorious pirate haven, where crews were fabled to live lives of great luxury and indulgence.

The success of pirates in the Indian Ocean eventually proved their undoing. To appease the powerful local rulers and merchants whose ships had been lost in the region, various European governments made suppression of piracy in the area a priority.

Famous pirates

Notable among the pirates who made their fortunes in the Indian Ocean were the English pirate Henry Every and the American-born Thomas Tew, who became overnight celebrities on their return, laden with rubies, silk, porcelain and other valuables. Their success encouraged other pirates to follow their example. However, less happy was the voyage of the infamous Captain Kidd, who sailed to the Indian Ocean to capture Every but actually ended up being hanged as a pirate himself.

FRENCH CORSAIRS

The word 'corsair' came from the French *la course*, meaning 'privateer'. The most famous corsairs hailed from the Breton port of St Malo, which grew rich on the successes of its privateering seamen. The tradition was so strong that it was common for sons to follow fathers into the trade, as if it was a family business.

DARING HEROES

The corsairs' daring deeds made them national heroes. The most celebrated was René Duguay-Trouin from St Malo who captured hundreds of ships during his 23-year career in the early years of the eighteenth century. Another hero was Robert Surcouf, who seized numerous English merchantmen in the Indian Ocean during the French Revolutionary and Napoleonic Wars. Also renowned was Jean Bart, from Dunkirk, who preyed on shipping in the English Channel in the second half of the seventeenth century.

The corsairs' activities infuriated the English, who were their usual victims, and in 1693 the English attempted to destroy St Malo by sailing a ship loaded with explosives into the port, only for it to detonate with very little effect, killing only one cat.

AMERICAN PRIVATEERS

War with France and Spain in the 1740s encouraged the issue of hundreds of letters of marque by the British colonial government authorizing American privateers to attack enemy shipping throughout the Caribbean and Atlantic.

THE AMERICAN REVOLUTION

When the American Revolution broke out in 1775 the rebels recruited over 400 privateering vessels to fight alongside the official American Continental Navy of just 34 ships. Despite the Royal Navy's dominance, the sheer numbers of American privateers damaged British trade and prompted Britain to make accusations of piracy against the privateers and the Continental Navy, whose heroes included John Paul Jones. More

American heroes

An American hero was the privateer captain Jonathan Haraden, who once called on a British vessel to surrender immediately or to face the massed fire of his cannon. The ship surrendered, unaware that Haraden had but a single shot left.

than 3,000 British ships and cargoes were seized by American privateers before the war ended in 1783.

Many privateers fought for patriotic reasons, passing captured cargoes of food, muskets and other supplies to the Revolutionary armies. Their only source of income, however, was prize money and thus personal enrichment was an important factor. Some of the cargoes they seized were very valuable, consisting of gold, ivory and other costly items, and several privateers made considerable fortunes.

THE WAR OF 1812

America called on the services of the privateers against Britain once more, in the War of 1812. Equipped with fast, well-armed, purpose-built long-range vessels, they seized more than 1,300 British prizes. The most famous of the privateers during the three-year campaign was Jean Lafitte, a pirate and slave-smuggler who became a national hero after defending New Orleans from British attack.

Most privateers retired with the restoration of peace, but the seas around the Americas continued to be infested with pirates, chiefly from Latin America, until the 1820s, when the British and American navies combined to clear the seas of the threat.

SOUTH CHINA SEA PIRATES

The seas of south-east Asia had a notorious reputation for piracy as early as the fifth century AD. Over the centuries, many pirate dynasties in the area rose and fell, sometimes wielding greater power than the imperial house of China itself. Whenever imperial control at sea faltered the pirates would quickly take advantage, attacking trading vessels of all nations and descriptions, and extracting protection money from coastal communities.

Organised piracy was restricted with the emergence of the Ming dynasty in the fifteenth century, chiefly through the rewarding of local rulers who took action against the pirate fleets. The development of trade between Europe and the Far East from the sixteenth century, however, meant that the potential rewards of piracy were greater than ever and rival pirate empires sprang up all along the Chinese coastline.

PIRATE EMPIRES

In the seventeenth century, Ching-Chi-ling wielded immense political power as the head of the greatest pirate empire yet established in the South China Sea. The empire was inherited by his son Kuo Hsing Yeh,

Pirate territory

The coastline of the South China Sea proved ideal pirate territory. Extensive mangrove swamps provided countless hiding places for pirate junks and made suppression of piracy doubly difficult.

who similarly exerted enormous political influence through his fleets of hundreds of pirate junks. The pirate empires declined in the early eighteenth century but revived on relocation to Vietnam.

Vast pirate fleets terrorized the region under the Tay Son dynasty and then under Cheng I, who controlled the largest pirate confederation ever assembled. When he died his empire was further expanded by his widow Cheng I Sao. Other powerful and greatly feared pirate chiefs included Shap'n'tzai and his lieutenant Chui Apoo, who were among the last of the great pirate leaders.

Piracy in the South China Sea was eventually suppressed in the mid-nineteenth century by a combination of bribes from the Chinese emperors and pressure from superior European warships protecting profitable trade routes to the area.

INDONESIAN PIRATES

The pirates of Indonesia were among the last to be suppressed. Operating out of hundreds of remote anchorages throughout the Indonesian archipelago, these pirates were organized along tribal lines and preyed chiefly on shipping between the Far East and Europe. For centuries the islands of Indonesia lacked any centralized authority and many local tribal kingdoms relied upon piracy for their income.

With the development of trade with Europe from the eighteenth century, the pirates of Indonesia flourished by attacking merchant vessels using the shipping lanes that passed through the region, especially those that had to sail through the narrow strait between Malaya and Sumatra. They often attacked in large canoes rowed by slaves, or else in fast sailing vessels with outriggers called *corocoros*, which might have crews as large as 60 in number.

The pirates of Indonesia, especially the Ilanun pirates of the Philippines, the Balanini pirates of Sulu, and the Sea Dyak pirates of Borneo, had a particularly fearsome reputation and many tales were told of the atrocities committed against the seafarers who were unlucky enough to fall into their hands. The captives were often murdered or otherwise mutilated and then cast adrift at the mercy of the waves.

SUPPRESSION BY THE NAVY

The establishment of the British colony of Singapore in 1819 paved the way for the restoration of law and order at sea. The Sea Dyaks and the Malay pirates were subdued by ships of the Royal Navy and the East India Company in 1836 and saw their bases destroyed, while the Ilanun and Balanini pirates were similarly overwhelmed by British and Spanish naval expeditions in the 1860s.

Here, Chinese pirates, who are brandishing swords, are preparing to attack a merchant ship.

PART TWO

Infamous pirates

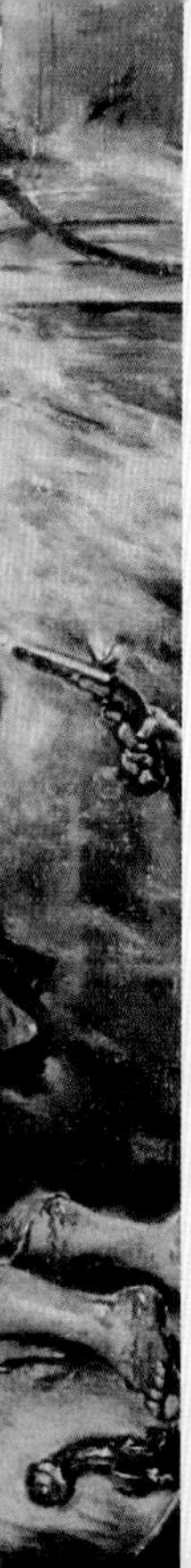

Among the many memorable characters to have entered pirate lore over the course of the centuries, a few have acquired almost legendary status and have been immortalized in fiction and film. Some have even touched the history of nations through their daring, cunning and ambition. Others, however, have claimed a lasting place in the popular imagination as monuments to greed and brutality.

Blackbeard in action
The pirate Edward Teach, more commonly known as 'Blackbeard', was one of the most fearsome pirates who ever lived. He is reputed to have worn lighted lengths of cord in his hair and drunk rum laced with gunpowder.

BARBAROSSA BROTHERS

BARBARY CORSAIRS

BORN: Mitylene, Greece, c.1470. **PIRATE CAREER:** c.1500–46. **WHERE ACTIVE:** western Mediterranean. **PRIZES TAKEN:** Spanish and other shipping. **DIED:** (Aruj) Killed in battle, Tlemcen, Algeria, 1518; (Hizir) 1546.

The brothers Aruj and Hizir Barbarossa were corsairs on the Barbary coast of northern Africa. The sons of a Moorish soldier, they sailed out of Alexandria before moving their galleys to Djerba, south of Tunis. From there they captured many prizes, including merchant vessels, warships and papal galleys. They moved their fleet to Djidjelli near Algiers in 1511.

ENEMIES OF THE SPANISH

In 1512 Aruj lost an arm attacking a Spanish fort and subsequently the brothers seized Spanish ships and settlements throughout the western Mediterranean. In 1516 Aruj led a corsair revolt against the Sultan of Algiers and claimed the sultanate for himself. The brothers repulsed a Spanish attack in 1518, but Aruj was killed in a siege soon after. Hizir (or Kheir-ed-din, meaning 'gift of God') assumed command and in 1530 became Sultan of Algiers after forming an alliance

The Barbary corsair Hizir Barbarossa was a wealthy pirate who worked with the Turks to consolidate the Ottoman Empire in the Mediterranean region.

with the Ottoman emperor. He continued to attack Spanish and other Christian targets, strengthening Ottoman control of the Mediterranean.

SIR JOHN HAWKINS

ENGLISH PRIVATEER

BORN: Plymouth, England, 1532. **PIRATE CAREER:** 1562–95. **SHIP:** *Jesus of Lubeck*. **WHERE ACTIVE:** Caribbean. **PRIZES TAKEN:** Portuguese and Spanish shipping. **DIED:** At sea, off San Juan, Puerto Rico, 1595.

Sir John Hawkins was a national hero whose success inspired many others. The son of a sea merchant, he made a fortune trading in slaves with the New World, defying the Spanish monopoly in the region. With backing from Queen Elizabeth I, who gave him his ship, he combined slaving with piracy against Spain during several voyages to the Caribbean. In 1568, he and Francis Drake were attacked by the Spanish and lost most of their vessels and plunder. He later became treasurer of the navy and in 1588 sailed to fight the Spanish Armada. He returned to piracy in 1590 and set out on a last expedition in 1595.

A sullied reputation

Although he was knighted for defeating the Armada, Sir John Hawkins' reputation today is shadowed by his involvement in the slave trade.

SIR FRANCIS DRAKE

ENGLISH PRIVATEER

BORN: Plymouth, England, c.1540. **PIRATE CAREER:** 1572–96. **SHIP:** *Golden Hind*. **WHERE ACTIVE:** Caribbean and east Pacific. **PRIZES TAKEN:** Spanish towns and shipping. **DIED:** Of a fever at sea, 1596.

Sir Francis Drake was a farmer's son who became one of the most famous English seafarers to roam the Spanish Main, making a fortune from captured Spanish galleons. Drake nursed a deep hatred of the Spanish after he and his cousin John Hawkins were attacked by them in the Gulf of Mexico in 1568, narrowly escaping with their lives.

In 1572, he received a commission as a privateer from Elizabeth I and sailed to the Caribbean to attack Spanish shipping and towns. A second three-year

Drake's drum

A drum that is alleged to have belonged to Sir Francis Drake is preserved at Drake's home, Buckland Abbey, Plymouth. It is said to beat to an unseen hand to warn England of danger.

expedition followed in 1577, his prizes including the Spanish treasure galleon *Nuestra Senora de la Conception* (nicknamed *Cacafuego*). It is said to have taken Drake's crew four days to transfer all the treasure to their own ship.

THE QUEEN'S PIRATE

Drake's activities brought much wealth to the English Crown and he was knighted in 1581 by Elizabeth I, who called him 'her pirate'. A national hero, Drake returned to the Caribbean with his own fleet in 1585, attacking the port of Vigo in Spain before crossing the Atlantic to raid Spanish colonies in the New World, including Santa Domingo in Hispaniola – the Spanish capital in the New World – Cartagena in Venezuela, and St Augustine in Florida. He sailed to the Spanish Main for the last time in 1596, dying of a fever, and was buried at sea.

Because Drake attacked Spanish vessels in peacetime as well as in wartime he was considered a pirate by his enemies. However, at home in England he was regarded as a national hero, especially after the defeat of the Spanish Armada. He is still honoured today as the first Englishman to circumnavigate the world in the *Golden Hind* as well as the man who destroyed the Spanish Armada in 1588.

Sir Francis Drake was knighted by Elizabeth I in 1581 for his services to England.

FRANÇOIS L'OLONNAIS

FRENCH BUCCANEER

Born: Les Sables d'Olonne, Brittany, France, c.1635.
Pirate career: c.1653–67. **Where active:** Caribbean.
Prizes taken: Spanish shipping and coastal towns.
Died: Killed by cannibals, Panama, 1667.

Born Jean-David Nau but calling himself L'Olonnais after his birthplace, he sailed to the Caribbean in 1650 and worked as a servant on Martinique before falling in with buccaneers, quickly rising to captain. Provided with a ship by the French governor of Tortuga, he preyed on Spanish shipping throughout the region, earning the nickname *Fléau des Espagnols* ('Flail of the Spaniards') and becoming notorious for his inhuman brutality to prisoners. Such was L'Olonnais's reputation that his victims often preferred to die in battle against him rather than surrender.

HIS EXPLOITS

These included an attack on a port in northern Cuba, during which he seized a warship and murdered the entire crew except for one man who was left to warn the governor of Havana that any Spaniard they took would be killed. In July 1667, L'Olonnais led eight ships against the wealthy towns of Maracaibo and Gibraltar in Venezuela, murdering the inhabitants and taking much treasure. Loaded with riches, the crews returned to Tortuga, where they quickly spent their new wealth. Their next expedition took them to the Gulf of Honduras, where they sacked Puerto Caballos and committed further atrocities.

KILLED BY CANNIBALS

L'Olonnais then set out for the silver mines of San Pedro further inland. However, this was to be his last expedition as he and his 700 men were ambushed by cannibalistic Darien Indians. Most of the buccaneers were killed and eaten, and L'Olonnais himself was torn limb from limb, screaming terribly, and was then burned to ashes.

Inhuman brutality

L'Olonnais is remembered as one of the most bloodthirsty of all the pirates to sail the seven seas. He once cut the still-beating heart out of a Spanish prisoner with his cutlass and gnawed on it as a warning to other prisoners of their fate if they did not cooperate before stuffing it into the mouth of another captive.

ROCK BRAZILIANO

DUTCH BUCCANEER

Born: Groningen, Netherlands, c.1630. **Pirate career:** 1654–c.1673. **Where active:** Caribbean. **Prizes taken:** Spanish and other shipping. **Died:** Probably in Port Royal, Jamaica, c.1673.

Rock Braziliano (also known as Roche Brasiliano) was a notoriously cruel buccaneer who operated out of Port Royal, Jamaica. He was a privateer in Bahia, Brazil, before moving to Port Royal in 1654. He led a mutiny and adopted the life of a buccaneer, seizing richly-laden Spanish treasure ships. The Spanish eventually caught him and sent him to Spain, but he escaped and resumed his career, buying a new ship off his fellow-pirate François L'Olonnais and later sailing in company with Sir Henry Morgan among others.

Atrocities

Drunken and debauched, Braziliano would threaten to shoot anyone who did not drink with him. His many atrocities including roasting alive two farmers on wooden spits after they refused to hand over their pigs.

BARTOLOMEO EL PORTUGUES

PORTUGUESE BUCCANEER

Born: c.1630. **Pirate career:** 1650s. **Where active:** Caribbean. **Prizes taken:** Spanish shipping. **Died:** Details unknown.

Bartolomeo was a Portuguese sailor who turned to piracy after Jamaica was captured by the English in 1655. A daring and ingenious rogue, he was dogged by ill luck throughout his career. As the captain of a small four-gun vessel, he managed to capture a much larger Spanish treasure ship after a bloody battle and with it a considerable haul of treasure. Unfortunately, three more Spanish galleons then appeared and Bartolomeo was captured. He narrowly escaped hanging, reputedly swimming ashore using earthenware jars as floats.

AN UNLUCKY PIRATE

Reduced to sailing in a seagoing canoe, he was nearly recaptured and later had to abandon his vessel after running aground off Cuba. The rest of his career was spent in more violent but fruitless attacks on Spanish shipping. He never made much from his crimes and is reported to have died in poverty.

CHARLOTTE DE BERRY

ENGLISH PIRATE

Born: 1636. **Pirate career:** c.1655–65. **Ship:** *Trader.* **Where active:** Off Africa. **Prizes taken:** English and other shipping. **Died:** Details unknown.

Charlotte de Berry dressed in men's clothes to follow her husband into the English navy, calling herself his brother. After her husband was flogged to death, she was forced onto a ship bound for Africa, the captain having discovered that she was a woman. When the captain assaulted her, she led a mutiny, beheading him with her dagger. The mutineers took to piracy, seizing gold ships off Africa.

Legend claims that Charlotte married a wealthy Spaniard in her crew but her ship sank and they and several others drifted for days on a raft. Starving, they drew lots to decide who should die to feed the others and her second husband lost and was killed.

Fact or fiction?

Charlotte de Berry may be entirely fictional as no records exist of her before 1836.

SIR CHRISTOPHER MYNGS

ENGLISH BUCCANEER

BORN: Norfolk, England, 1625. **PIRATE CAREER:** 1656–65. **SHIPS:** *Marston Moor, HMS Centurion, HMS Victory*. **WHERE ACTIVE:** Caribbean. **PRIZES TAKEN:** Spanish and Dutch shipping and coastal settlements. **DIED:** Killed in battle, August 1666.

Sir Christopher Myngs was a naval officer, organizing English buccaneers against the Spanish in the Caribbean. He arrived at Port Royal in 1656 with orders to defend Jamaica. In 1658, his ships sacked towns in Colombia and Venezuela and captured a large shipment of Spanish silver. Because the booty was not shared equally with the Crown, Myngs was tried for embezzlement, but the case was dropped and by 1662 he was back in Jamaica where he continued to harass the Spanish, taking Santiago in Cuba and raiding the coastline of Mexico.

Rewarded for service

Myngs was made a vice-admiral on returning to England in 1665 but a year later was killed in the Second Dutch War.

SIR HENRY MORGAN

WELSH BUCCANEER AND PRIVATEER

BORN: Monmouthshire, Wales, c.1635. **PIRATE CAREER:** 1661–72. **SHIPS:** *Oxford, Marquesa, Satisfaction*. **WHERE ACTIVE:** Caribbean. **PRIZES TAKEN:** Spanish towns and shipping. **DIED:** Port Royal, Jamaica, 25 August 1688.

Henry Morgan may have served as a soldier in the Caribbean but by 1661 was preying on Spanish shipping with his own vessel and raiding their towns. He was a natural leader who insisted upon military discipline and showed little mercy to enemies.
In 1663, he received a privateering commission and assembled a small fleet. Over the next two years he sacked Villahermosa and Gran Granada, seizing booty. He also befriended the governor of Jamaica, Thomas Modyford, who encouraged him to attack Spanish interests in the Caribbean.

Sailing from Port Royal, Jamaica, in January 1688, Morgan took 10 ships to southern Cuba, where he was joined by French buccaneers from Tortuga. Together they marched 48 km (30 miles) inland to attack Puerto Príncipe, forcing the defenders to surrender. The population were locked inside their churches or tortured to reveal their treasure.

Sir Henry Morgan is shown here with booty and a female companion celebrating a successful raid.

Subsequently Morgan attacked Porto Bello by land, using prisoners as human shields. His men took the three forts and the city. Captives were tortured into revealing hidden treasure and Morgan refused to leave until a large ransom was paid. Charges of piracy against Modyford and Morgan were dropped and in October 1668 the latter joined forces with French buccaneers to attack Cartagena. Morgan's ship was destroyed in an accidental explosion, killing 200 crew and obliging him to attack Maracaibo and Gibraltar in Venezuela instead.

THE SACK OF PANAMA

Despite peace with Spain in 1669, in 1670 Modyford asked Morgan to assemble 2,000 buccaneers and sail with 36 ships to Panama, the richest prize in Spanish America. Having paddled up the Chagres River by canoe, the pirates marched through the jungle to the city, where they routed the defenders, killing 500. When little treasure was found, they destroyed the city and maltreated the population. This was the last great buccaneering venture. Modyford and Morgan were sent to England under arrest but Morgan was hailed a hero, knighted and made lieutenant-governor of Jamaica. He returned home, a wealthy plantation-owner and the most successful of all the buccaneers.

Keeping the booty

Morgan's commission dictated that any treasure taken through the use of ships had to be split with the English Crown. So he attacked Spanish towns from the land, retaining all the booty.

MICHEL DE GRAMMONT

FRENCH BUCCANEER

BORN: Paris, France, c.1645. **PIRATE CAREER:** c.1670–86. **SHIP:** *Hardi*. **WHERE ACTIVE:** Caribbean. **PRIZES TAKEN:** Spanish and Dutch shipping and coastal settlements. **DIED:** Lost at sea, north-east Caribbean, April 1686.

Michel de Grammont may have served in the French navy before emerging as a buccaneer commander, operating out of Tortuga and Saint Dominique against Spanish shipping. When war broke out between France and Holland in 1678, he took part in an abortive raid on the Dutch island of Curaçao, and was appointed commander of the buccaneers.

In June 1678 he led six ships and 700 men into Spanish-held Venezuela, penetrating as far inland as Trujillo. This was followed by another successful raid on the Venezuelan port of La Guaira, captured in a daring night attack, though the buccaneers only escaped with difficulty when attacked by a larger Spanish force. In May 1683, de Grammont and Dutch buccaneer Laurens de Graaf sacked Vera Cruz in Mexico, taking 4,000 prisoners for ransom. In 1685, the pair also sacked the Mexican city of Campeche and held it for three months.

JEAN BART

FRENCH CORSAIR

BORN: Dunkirk, France, 1651. **PIRATE CAREER:** 1673–1702. **SHIP:** *Le Roi David*. **WHERE ACTIVE:** English Channel. **PRIZES TAKEN:** English, Dutch, Spanish ships. **DIED:** 1702.

Jean Bart served in the Dutch navy, becoming a corsair when war broke out with Holland in 1672. A brilliant seaman, he made a fortune from prizes seized from France's enemies. His daring won him many admirers and in 1694 he was ennobled by Louis XIV and given command of a French fleet. Captured by the English, he escaped by rowing back to France in a small boat.

Jean Bart, in disagreement with his crew members, is portrayed threatening to blow up his ship.

BARTHOLOMEW SHARP

ENGLISH BUCCANEER

Born: Unknown, c.1650. **Pirate career:** 1679–82. **Ship:** *Trinidad*. **Where active:** South America. **Prizes taken:** Spanish shipping, coastal towns. **Died:** 1690.

Bartholomew Sharp's career as a pirate captain began when the buccaneers with whom he was sailing round South America needed a new commander. He quickly proved himself a natural leader and a capable seaman. These qualities did not prevent him being deposed as captain in January 1681, however, after storms and setbacks provoked a mutiny. His successor was killed three weeks later and Sharp resumed command. Under him the buccaneers continued around South America and up to the Caribbean, taking 25 Spanish ships and plundering numerous Spanish towns.

A ROYAL PARDON

Because England and Spain were not at war the Spaniards demanded Sharp's prosecution for piracy. Sharp, however, presented the authorities with a book of maps taken from the Spanish ship *El Santo Rosario* in July 1681; their value to English seafarers was such that Sharp received a full pardon from Charles II.

WILLIAM DAMPIER

ENGLISH BUCCANEER AND EXPLORER

BORN: Near Yeovil, Somerset, 1652. **PIRATE CAREER:** 1679–1715. **SHIPS:** *Royal Prince, Cygnet, HMS Roebuck, Saint George*. **WHERE ACTIVE:** Caribbean, Atlantic, Pacific, South China Sea. **PRIZES TAKEN:** Spanish towns, Spanish and other shipping. **DIED:** London, 1715.

William Dampier first went to sea aged 17, as a deck hand on a merchant ship. In 1673, he joined the Royal Navy, fighting the Dutch as a sailor on board the *Royal Prince*. Leaving the navy because of illness, he spent some time as a plantation manager in Jamaica and logwood cutter before joining the crew of buccaneer Bartholomew Sharp (see page 61) in 1679. Over the next three years, Sharp's crew seized many Spanish ships and raided Spanish towns along the coast of South America.

VOYAGES AROUND THE WORLD

In 1683, Dampier sailed with buccaneer Captain John Cook in search of prizes on both sides of the Atlantic and in the Pacific. Two years later he sailed with Captain Swan to the East Indies and explored the coasts of China and Australia (then called New Holland).

The explorer William Dampier and his crew were attacked by Aborigines in Australia.

Dampier's journals

William Dampier recorded his various voyages in a celebrated series of journals, which present a colourful picture of a seafaring life. Among other things, he describes how he ate and enjoyed many exotic foods on his travels, his favourite dishes including flamingo's tongue.

Dampier completed his second circumnavigation in the early 1700s, this time as captain of *HMS Roebuck*. The voyage was not a happy one, however, with Dampier falling out with his first lieutenant and the ship itself eventually sinking. Dampier was court-martialled and declared unfit to command a King's ship.

In disgrace

Nonetheless, in 1703 Dampier was given command of two privateers on an expedition to the South Seas. This too ended badly, with Dampier being accused of financial greed, incompetence, drunkenness, brutality and cowardice. In order to escape criticism from the expedition's backers, the impoverished and broken Dampier immediately left on another voyage, acting as pilot to the West Indies for the privateer Captain Woodes Rogers. He died back in London four years after this voyage, aged 63.

RENÉ DUGUAY-TROUIN

FRENCH CORSAIR

BORN: St Malo, France, 1673. **PIRATE CAREER:** 1689–1712. **SHIPS:** *Trinité, Danycan, Hermine, Diligente, Bellone, Railleuse.* **WHERE ACTIVE:** English Channel, Atlantic. **PRIZES TAKEN:** English and other shipping. **DIED:** 1736.

René Duguay-Trouin joined the French navy aged 16. He was given his first ship at 18 and the 40-gun *Hermine* at 21. He amassed a fortune as a corsair at the cost of France's enemies, especially the English and Dutch. He was captured by the English in 1694 but escaped and was presented to the French king Louis XIV in 1695. He was ennobled in 1709 and, in 1711, during the War of the Spanish Succession, captured the supposedly impregnable Rio de Janeiro after an 11-day bombardment, forcing the city to pay a ransom. The most celebrated of all French corsairs, Duguay-Trouin rose to the rank of admiral in the French navy.

A prolific career

Brave and respected by his men, Duguay-Trouin seized 20 warships or privateers and some 300 merchant vessels during his career.

THOMAS TEW

AMERICAN PIRATE

Born: Newport, Rhode Island, c.1660. **Pirate career:** 1692–95. **Ship:** *Amity*. **Where active:** Indian Ocean and Red Sea. **Prizes taken:** Indian and Arab shipping. **Died:** Killed at sea, September 1695.

Thomas Tew is illustrated in the company of his friend, the English governor of New York.

Thomas Tew, the son of a respectable Rhode Island family, sailed from Bermuda in the eight-gun ship *Amity* planning to make his fortune from piracy in the Indian Ocean.

In 1694, he returned to Newport, Rhode Island, with a fabulous haul of treasure seized from a ship belonging to the Great Moghul of India and he became an instant celebrity.

KILLED AT SEA

The tales of Tew's adventures circulated widely and inspired many others to plan similar piratical exploits. Tew embarked on a second voyage in November 1694 and may even have sailed in company with fellow-pirate Henry Every before being killed in battle while he was attempting to capture the Indian merchantman *Fateh Mohammed*. His crew were captured and probably executed.

English backing

At the height of his fame, Tew's friends included no less than the English governor of New York, who found Tew 'agreeable and companionable' and backed his second – fatal – voyage.

HENRY EVERY ('Long Ben')

ENGLISH PIRATE

Born: Plymouth, England, c.1653. **Pirate career:** 1694–96. **Ship:** *Fancy*. **Where active:** Indian Ocean. **Prizes taken:** English, French, Arab shipping. **Died:** c.1728.

Henry Every (or Avery) worked in the slave trade before turning to piracy. Early in 1694 he joined the crew of a 46-gun privateer but subsequently led a mutiny and took the ship to the Indian Ocean to prey on shipping.

THE MOGHUL'S TREASURE

By 1695, Every was the leader of a formidable pirate fleet of five vessels. In the Red Sea, he captured two treasure ships belonging to the Moghul of India, the *Fateh Mohammed* and the *Gang-i-Sawai*. The crews were brutally treated (although Every took no part in this) and tortured into revealing the whereabouts of the treasure. Everyone in the pirate fleet became immensely wealthy and Every retired, becoming a folk hero. Legend has it that he settled in Madagascar with his captive, the Moghul's beautiful daughter. But there are claims that he died penniless in a Devon cottage, without funds even for a coffin. Many of his crew were hanged after returning to England.

WILLIAM KIDD ('Captain Kidd')

SCOTTISH PRIVATEER

Born: Greenock, Strathclyde, c.1645. **Pirate career:** 1697–1700. **Ships:** *Blessed William, Antegoa, Adventure Galley, Adventure Prize, Antonio.* **Where active:** Indian Ocean. **Prizes taken:** English, Portuguese shipping. **Died:** Hanged, Wapping, London, 1701.

William Kidd ranks among the most infamous of pirates despite the fact that he never believed himself to be one in the strictest sense. He spent the 1680s as a wealthy and respected captain with a small trading fleet, operating out of New York. Subsequently, he adopted the role of a privateer protecting Anglo-American trade routes in the Caribbean against France. He returned to London in 1695 and was persuaded by the Earl of Bellamont and other political backers to lead a legally dubious expedition to rid the Indian Ocean of pirates, the aim of which was largely financial gain.

As captain of the 34-gun *Adventure Galley*, Kidd reached Madagascar in 1697 but there most of his crew, objecting to the fact that neither they nor their captain stood to gain much financially, demanded that they become pirates themselves. From that point on, Kidd sanctioned attacks on any kind of shipping off the coast of India. Having put down a possible mutiny by killing one of the ringleaders, he quickly amassed a substantial hoard of treasure, his greatest prize being the 400-ton treasure ship *Quedah Merchant*, which he renamed the *Adventure Prize* and made his new flagship.

A POLITICAL PAWN

Kidd returned to the Caribbean, where he discovered he had been branded a pirate. He sailed to Boston, believing he could rely upon the protection of Lord Bellamont, governor of New York and Massachusetts, and maintaining that he had attacked only French and pirate vessels. Unfortunately for Kidd, Bellamont (conscious of potential political embarrassment) refused a pardon and instead had Kidd clapped in irons and sent to London, where he was tried for murder and piracy and hanged.

Kidd is said to have buried his treasure on Long Island, outside New York. Legend claims that the pirates drew lots to decide which of them would be killed so that his body could be left on top of the treasure chest to warn off treasure hunters. It is probable that Governor Bellamont retrieved the treasure after Kidd's arrest.

EDWARD TEACH ('Blackbeard')

ENGLISH PIRATE

BORN: Bristol, date unknown. **PIRATE CAREER:** 1716–18. **SHIPS:** *Queen Anne's Revenge, Adventure*. **WHERE ACTIVE:** Caribbean, east coast of America. **PRIZES TAKEN:** French and other shipping. **DIED:** Killed, Ocracoke Inlet, off North Carolina, 22 November 1718.

Edward Teach, Thatch or Tash (he was possibly born Edward Drummond) was the most infamous pirate of them all and his exploits are well recorded and have captured the popular imagination. Nicknamed Blackbeard after his distinctive long black beard, he was tall and broad-shouldered, and carried two swords, six pistols and knives.

TERRIFYING APPEARANCE

Cunning, ruthless and unpredictable, Edward Teach inspired great fear and even some of his own crew believed that with his wild, staring eyes and cruel ways he was the Devil himself. To heighten the frightening effect still further, he took to wearing lighted lengths of cord in his hair and twisted his beard into long black plaits. His favourite drink was rum laced with gunpowder.

The infamous pirate Edward Teach (Blackbeard) resisting arrest by Robert Maynard in 1718.

EARLY CAREER

Teach began his career under the pirate Benjamin Hornigold. In 1716, Hornigold gave him a small ship and together, operating out of Jamaica, they spread terror throughout the Caribbean. In November of the same year Hornigold retired and Teach converted the captured French slaver *Concorde* into the 40-gun *Queen Anne's Revenge* .

Attacks on the American coast

Shortly afterwards Teach encountered fellow-pirate Stede Bonnet with his 10-gun sloop *Revenge*. He appointed a man to take over the *Revenge* and forced Bonnet to live on the *Queen Anne's Revenge* as his guest. He then rampaged up the American coast, taking many prizes and even blockading Charleston harbour and kidnapping prominent local citizens.

By spring 1718, he had amassed four ships and 300 crew. Indeed, such was Teach's reputation that many of his victims surrendered without a fight, knowing that if they resisted he would kill or maroon them.

In May 1718, the *Queen Anne's Revenge* finally ran aground and therefore Edward Teach transferred to the 10-gun *Adventure*, leaving Stede Bonnet to go his own way.

Teach's women

Many myths surround Edward Teach. He is said, for instance, to have had 14 wives. One story claims that he was once rejected by a pretty girl in favour of another seaman, to whom the girl gave a ring. However, when Teach later captured the man's ship he recognized the ring. He subsequently cut off the lover's hand and then sent it to the girl who had spurned him in a silver casket. She died of grief.

BLACKBEARD'S LAST BATTLE

Justice caught up with Teach on 22 November 1718 when his ship was trapped in the pirate hideout Ocracoke Inlet by two Royal Navy sloops. Teach and his men were tricked into boarding Lieutenant Robert Maynard's sloop *Jane*, only to find themselves surrounded. They fought desperately, Teach himself battling ferociously with Maynard, using both pistol and cutlass. He sustained five bullet wounds and more than 20 sword cuts before he died. Maynard sailed home with Teach's severed head hanging from the bowsprit at the front of his ship. Legend records that Teach's headless body ran wildly around the deck before throwing itself into the sea.

CHARLES VANE

ENGLISH PIRATE

Born: Details unknown, c.1680. **Pirate career:** 1716–20. **Ship:** *Ranger*. **Where active:** Caribbean and off the east coast of north America. **Prizes taken:** English, French, and other shipping. **Died:** Hanged, Gallows Point, Port Royal, Jamaica, 29 March 1720.

Charles Vane was among the pirate captains who established a notorious base at New Providence in the Bahamas after the British abandoned the colony in 1713. When threatened there in August 1718 by Governor Woodes Rogers and two Royal Navy ships, Vane alone resisted them, driving the men-of-war back with a captured French fireship. Vane then escaped in his fast six-gun sloop, the *Ranger*, defiantly firing on the governor as he passed and threatening to return.

A cruel cheat

The pirate Charles Vane was despised for his cruelty. He also showed scant respect for the pirate code, cheating his own crews out of their fair share of plunder.

DEPOSED FOR COWARDICE

Vane subsequently traded up ships by capturing first a Barbados sloop and then a large 12-gun brigantine, which he also renamed the *Ranger*. He evaded his Royal Navy pursuers and in October 1718 even enjoyed a week-long celebration at Ocracoke Inlet, North Carolina, with Blackbeard and his crew.

Vane then cruised north to New York, seizing yet more vessels, before turning south towards the Caribbean, only for his crew to vote him out of his captaincy for cowardice after failing to engage a larger French warship. Replaced by his quartermaster 'Calico Jack' Rackham, he was cast adrift in a small sloop. Subsequently he set about clawing his way back up the pirate ranks by seizing ever larger ships.

SHIPWRECKED

Vane's final downfall came after his ship was wrecked in a hurricane in February 1719 and he and one other survivor were washed up on an uninhabited island in the Bay of Honduras. A ship eventually arrived, months later, but Vane was recognized and both men were clapped in irons, taken to Jamaica, and there tried and hanged. Vane died without expressing the least remorse for his crimes.

STEDE BONNET

ENGLISH PIRATE

Born: Barbados, 1689. **Pirate career:** 1717–18. **Ship:** *Revenge* (renamed the *Royal James*). **Where active:** Off the east coast of north America. **Prizes taken:** English and Spanish shipping. **Died:** Hanged, Charleston, Carolina, 12 November 1718.

Stede Bonnet was an English gentleman and former army major who owned a plantation in Barbados before achieving fame as 'the gentleman pirate'. Unusually, he is said to have used his own money to buy his ten-gun sloop *Revenge*.

With a volunteer crew he plundered shipping along the coast from New York to the Carolinas. However, unfortunately for Bonnet, he then ran into the feared Edward Teach (Blackbeard), who assumed command

Walking the plank

Though Bonnet dressed like a gentleman and was evidently well-educated, he had a ruthless streak and was one of the few pirates said to make prisoners walk the plank.

of the less experienced pirate's ship and crew. Bonnet became a virtual prisoner and, it is said, was even put to work as Blackbeard's servant.

LIFE AS A PRIVATEER

After eventually getting his ship back, Bonnet learned that Great Britain was at war with Spain and offered his services to the governor of North Carolina (then owned by the British Crown) in exchange for a pardon. He then operated as a privateer, targeting Spanish vessels and pursuing his old enemy Blackbeard. He was soon back to his piratical ways, however, disguising his identity by calling himself Captain Thomas and his ship the *Royal James*.

END OF THE GENTLEMAN PIRATE

When Bonnet put into the Cape Fear river for repairs he was attacked by two ships sent by the authorities in Charleston, South Carolina, under the command of the pirate-hunter William Rhett. After a five-hour fight the pirates were eventually taken prisoner. Bonnet escaped briefly but he was recaptured, tried, and sentenced to death, together with 30 of his crew. Although he expected a reprieve on the basis of his gentlemanly status, he was duly hanged in Charleston harbour.

EDWARD ENGLAND

IRISH PIRATE

Born: Ireland, c.1690. **Pirate career:** 1717–20. **Ships:** *Pearl* (renamed the *Royal James*), *Fancy*. **Where active:** Caribbean, east Atlantic, Indian Ocean. **Prizes taken:** English and other shipping. **Died:** Madagascar, 1720.

Edward England operated out of New Providence in the Bahamas before the Caribbean was cleared by the Royal Navy in 1718. He then transferred his activities to the seas off Africa, operating out of Madagascar.

In 1720 he and Captain John Taylor fought the East Indiaman *Cassandra* off Madagascar, capturing the ship and £75,000 of valuables. Taylor favoured killing James McRae, its captain, but England let him go and was deposed as captain and marooned on Mauritius. He built a raft and drifted 500 miles to Madagascar.

An ignominious end

England was among the many pirates whose lives ultimately were fated to end in wretched circumstances, in his case dying a penniless beggar in Madagascar.

HOWELL DAVIS

WELSH PIRATE

BORN: Milford Haven, Monmouthshire, Wales, c.1700. **PIRATE CAREER:** July 1718–June 1719. **SHIPS:** *Cadogan, Buck, Saint James, Rover*. **WHERE ACTIVE:** Caribbean, West African slave coast. **PRIZES TAKEN:** French, Portuguese, Dutch, and English shipping and Gambian slaving fort. **DIED:** Shot, Principé, June 1719.

Howell Davis started out in piracy after the slave ship *Cadogan* on which he was serving with the rank of ship's mate was captured by the pirate ship of Edward England in 1718. Having agreed to join the pirates, Davis was given the *Cadogan* and set out for Brazil but his crew mutinied and sailed to Barbados instead. Here Davis was imprisoned on charges of piracy, but was eventually released and sought shelter in the pirate den of New Providence in the Bahamas. With New Providence being cleaned up by Governor Woodes Rogers, Davis left on the sloop *Buck* and, after persuading the crew to take over the vessel off Martinique, was elected captain and conducted piratical raids from his base at Coxon's Hole on Cuba.

Subsequently he crossed the Atlantic to terrorize shipping in the Cape Verde Islands. One of the prizes

A dashing and charming figure, Howell Davis is remembered for his cunning and deceit.

taken there became the new flagship of Davis' pirate fleet, the 26-gun *Saint James.*

A DECEITFUL CHARMER

A clever and charming man, Davis pretended to be a legitimate privateer to deceive the commander of a Royal African Company slaving fort in Gambia and held him to ransom after capturing him at a welcome dinner. He then formed a partnership with a French pirate known as La Bouche, which lasted until they fell out in a drunken argument. Transferring to the 32-gun *Rover*, he sailed south and captured more rich prizes off the Gold Coast, his prisoners including fellow-Welshman Bartholomew Roberts, who was destined to become equally famous as a pirate.

When Davis tried to repeat his earlier trick of pretending to be a privateer, for the benefit of the governor of the Portuguese island of Principé (whom he planned to kidnap), he was unmasked and killed in an ambush by Portuguese militia.

Davis was famed for his use of trickery and deception. He once seized a more powerful French vessel by flying a black pirate flag from another large but only lightly armed ship he had recently taken: the French ship quickly surrendered, thinking it was outgunned.

BARTHOLOMEW ROBERTS

WELSH PIRATE

Born: South Wales, 1682. **Pirate career:** 1719–22. **Ships:** *Royal Rover, Royal Fortune*. **Where active:** Both sides of the Atlantic. **Prizes taken:** Portuguese, French, English and Dutch shipping. **Died:** Killed in action off Cape Lopez, Gabon, 10 February 1722.

Bartholomew Roberts (born John Roberts) is often described as the most successful pirate of the 'Golden Age' of piracy. During his brief but colourful career he took more than 400 prizes and amassed a fortune.

'Black Bart' was serving as ship's mate on a slave ship when it was captured by the pirate Howell Davis in June 1719. Roberts joined the pirates and when Davis was killed shortly afterwards was elected his successor. Over the next six months he took some 200 ships off the Americas, including several from a Portuguese convoy off Brazil.

He made a 28-gun French vessel his new flagship, renaming it the *Royal Fortune*, and took another 15 ships in the Caribbean during the summer of 1720. Another 100 ships fell to him as he cruised the West Indies and he also attacked St Kitts and Martinique.

A ruthless pirate

Although good-looking, well-dressed and teetotal, Roberts was ruthless and deserved his nickname of 'Black Bart'. For instance, when the owner of a captured slave ship refused to pay a ransom, he had it burned, along with its cargo of slaves.

His prisoners included the governor of Martinique, whom Roberts hanged from the yardarm of his own 52-gun warship. The crew were murdered and the ship itself became Roberts' new flagship, also renamed the *Royal Fortune*.

THE DEATH OF BLACK BART

Roberts operated off west Africa from April 1721, seizing several slave ships, of which one became his next flagship (again renamed the *Royal Fortune*). His luck ran out, however, on 10 February 1722 when he was caught at anchor by *HMS Swallow* off modern Gabon. Roberts died in a broadside of grapeshot fired at close range. The rest of the crew surrendered after a three-hour battle, but not before they had tossed their captain's body overboard so it would not be captured. Of the surviving crew, 54 were later hanged.

ANNE BONNY & MARY READ

IRISH/ENGLISH PIRATES

BORN: (Bonny) County Cork, Ireland, c.1699; (Read) London, c.1699. **PIRATE CAREER:** 1719–20. **SHIP:** *William*. **WHERE ACTIVE:** Caribbean. **PRIZES TAKEN:** Spanish shipping. **DIED:** (Bonny) Port Royal, Jamaica, 1720; (Read) details unknown.

Anne Bonny and Mary Read (or Reade) were the most famous of the women pirates. Anne Bonny was the wife of a sailor, but deserted him for dashing pirate captain 'Calico' Jack Rackham (or Rackam) in New Providence in 1718, assuming loose-fitting male dress in order to live the life of a pirate with him (women not being allowed on pirate ships). Mary Read first dressed as a male as a child and fought as a man in both the army and navy, but became a pirate after her ship was seized by Rackham.

Courageous women

Both women were respected by their fellow-pirates for their courage. They took part in boarding parties, fought duels and generally did everything any full-blooded pirate did.

The female pirate Mary Read bares her breast to a shocked combatant, revealing her sex.

The story goes that Bonny fell in love with Read, not realizing that she was a woman. Read then revealed her secret to Bonny and the two became friends.

CAPTURE OF THE WOMEN PIRATES

When Rackham's drunken crew were surprised by a Royal Navy sloop off Jamaica, Bonny and Read alone resisted arrest, the rest hiding in the hold. All were captured, but the women escaped the death penalty as both were expecting babies. As Rackham was hanged in Port Royal, Jamaica, Bonny told him: 'Had you fought like a man, you need not have been hanged like a dog!' Read died of a fever in prison soon afterwards. Bonny's fate is unrecorded.

It remains unclear whether Rackham's crew knew the headstrong, foul-mouthed pair's secret. Some say that the truth only came out later at their sensational trial. Others claim, however, that they only wore men's clothes when fighting.

Astonishing claims

Legend claims Read won a duel with another pirate by suddenly baring her breasts and then decapitating him as he stared in astonishment.

Anne Bonny, dressed here in men's clothes, was a friend and associate of Mary Read (see page 85).

EDWARD LOW

ENGLISH PIRATE

BORN: London, c.1690. **PIRATE CAREER:** 1721–24. **SHIPS:** *Fancy, Rose, Merry Christmas, Fortune.* **WHERE ACTIVE:** East coast of north America, Caribbean, Azores. **PRIZES TAKEN:** English, French, Portuguese, and Spanish shipping. **DIED:** Details unknown.

Edward Low earned a reputation for sadistic cruelty. His many atrocities included cutting off a man's lips and frying them in front of him before murdering him and his crew. He also severed the ears of a captive and made him eat them. Such excesses provoked a mutiny in 1724 and Low was set adrift in a small boat. He was rescued and resumed his career. When his two ships were attacked by *HMS Greyhound* east of Long Island, he sailed away, leaving the crew to be taken prisoner, and 26 of them were hanged.

A sentimental man

Despite his bloodcurdling reputation, Low often wept over the death of his young wife and the son he left behind in Boston. Perhaps because of this he had only unmarried men in his crew.

JOHN PAUL JONES

SCOTTISH-BORN AMERICAN NAVAL OFFICER

BORN: Kirkbean, Kirkcudbrightshire, Scotland, 6 July 1747. **PIRATE CAREER:** 1775–81. **SHIPS:** *Providence, Ranger, Bonhomme Richard*. **WHERE ACTIVE:** British coastline and the North Sea. **PRIZES TAKEN:** British shipping. **DIED:** Paris, 18 July 1792.

John Paul Jones first served as an apprentice on a merchant vessel and as ship's mate on a slaving ship but abandoned trade in the Caribbean after being accused of murder. In 1775, he joined the American navy as a lieutenant and emerged as a hero of the War of Independence for his daring actions against British shipping, although he was condemned as a pirate by his enemies. As captain of the 18-gun *Ranger*, he even conducted raids on the coast of Britain itself, attacking the harbour at Whitehaven and making landings in Scotland.

A FAMOUS ENGAGEMENT

Particularly celebrated was the engagement between Jones's 40-gun frigate *Bonhomme Richard* and the larger British warship *Serapis* off Flamborough Head, during which Jones's ship appeared doomed. When called on to surrender, however, he replied: 'I have not yet begun to fight!' Three hours later it was the *Serapis* that struck its colours.

Captain John Paul Jones is depicted here shooting a sailor who had attempted to strike his colours in an engagement at sea.

ROBERT SURCOUF

FRENCH CORSAIR

Born: St Malo, Brittany, France, 1773. **Pirate career:** 1795–1809. **Ship:** *La Confiance*. **Where active:** Indian Ocean. **Prizes taken:** British shipping. **Died:** 1827.

Robert Surcouf was one of the most celebrated seafarers to emerge from the famous French corsair base of St Malo in Brittany. Sailing to the Indian Ocean, he operated out of the island of Mauritius, his prizes including many British merchant ships bound for Indian ports. His greatest triumph was the capture after a fierce battle of the 38-gun British East Indiaman *Kent* as captain of the much smaller 18-gun *La Confiance*. Returning to France with a great haul of treasure, Surcouf was made a baron by the emperor Napoleon and subsequently established St Malo and other French ports as havens for French privateers. He retired from the sea at the age of 36.

A brave corsair

Robert Surcouf was renowned for his personal bravery, once even defeating a dozen Prussian soldiers single-handed.

JEAN LAFITTE

FRENCH-BORN AMERICAN PIRATE AND SMUGGLER

Born: Haiti or Bayonne, France, c.1780. **Pirate career:** c.1800–26. **Where active:** Indian Ocean, Gulf of Mexico. **Prizes taken:** British, American and Spanish shipping. **Died:** Details unknown, c.1826.

Little is known about the early career of Jean Lafitte, but by 1807 he and his brother Pierre were active as pirates in Barataria Bay, south of New Orleans, and running a huge smuggling operation in the city itself, where they employed one in ten men. Particularly lucrative was their involvement in the slave trade (though slaving was illegal in Louisiana and their auctions had to be held in secret). Lafitte was daring and dashing, but this did not prevent the arrest of the brothers in November 1812 on charges of piracy and illegal trading. Undeterred, however, they managed to escape and return to their piratical careers.

THE BATTLE OF NEW ORLEANS

During the War of 1812 against Britain the British tried to bribe Lafitte to help them capture New Orleans. Lafitte warned the authorities, only for his

Rewards in kind

A local governor once offered $500 for the capture of Lafitte. The pirate responded by offering $5,000 for the arrest of the governor.

pirate fleet to be seized by the American schooners *Carolina* and *Louisiana*. In December, however, the newly-arrived General Andrew Jackson accepted his offer of help. When the British attacked the city on 8 January 1815 Lafitte and his men manned guns landed from the two American warships and helped drive the attackers away.

A PIRATE DEN

The brothers were granted an official pardon but they were soon back to their old ways. When they were forced out of Barataria Bay, they stole a ship and went on to establish a new base at Galveston, Texas (then Spanish-held). Galveston quickly became a notorious pirate den, home to 20 pirate ships and with its own slave market. Attacks on American shipping prompted the authorities to destroy Galveston in 1821.

The fate of Lafitte is obscure: he may have died in Mexico, been killed or lived quietly on under an alias.

CHENG I SAO

CHINESE PIRATE

Born: Details unknown, 1775. **Pirate career:** 1807–10. **Where active:** South China Sea. **Prizes taken:** Chinese and European shipping. **Died:** Canton, 1844.

Cheng I Sao (or Ching Shih) worked as a prostitute in Canton until she married the powerful pirate Cheng I (or Ching Yih), leader of a huge pirate confederation. After his death, Cheng I Sao inherited command and expanded her empire until she had around 1,800 armed junks and 80,000 pirates under her control. This great fleet of ships spread terror throughout the China Sea and dominated China's trade routes. Under pressure from British and Portuguese warships and bribes from the Chinese court, the confederation finally broke up in 1810. Cheng I Sao herself accepted an imperial pardon, retiring to Canton.

A powerful pirate queen

Arguably the most powerful pirate commander since classical times, pirate queen Cheng I Sao was even beyond the control of the Chinese emperor himself.

BENITO DE SOTO

PORTUGUESE-BORN PIRATE

BORN: Portugal, c.1800. **PIRATE CAREER:** 1827–32. **SHIP:** *Black Joke*. **WHERE ACTIVE:** Caribbean and Atlantic. **PRIZES TAKEN:** Spanish and other shipping. **DIED:** Hanged, Cadiz, Spain, 1832.

Benito de Soto was the most notorious of the last generation of pirates to plunder shipping in the Atlantic. De Soto served on an Argentinian slave ship before leading a mutiny off south-west Africa in 1827. When 18 of the crew declined to participate they were cast adrift in an open boat and eventually drowned. De Soto then murdered the mate and was elected captain.

Having renamed the vessel the *Black Joke*, de Soto crossed the Atlantic, where he sold the stolen cargo of slaves, and then sailed south, attacking Spanish

A black joke

When a captured seaman helped pilot the *Black Joke* into a Spanish harbour de Soto thanked him – and then blew his brains out.

ships along the east coast of South America. De Soto proved to be one of the most bloodthirsty pirates of any age, murdering crews who fell into his hands and sinking their ships. The *Black Joke* then ventured eastwards into the Atlantic to intercept vessels returning from India and the Far East.

BRUTAL ATROCITY

The most infamous episode in de Soto's career came on 21 February 1832, when the *Black Joke* happened upon the *Morning Star* en route from Ceylon to England. After killing some of the passengers and crew with cannon-fire, de Soto murdered the captain with his cutlass and took possession of the ship.

Many of the captured crew were killed, while women passengers were raped before being locked in the hold with the rest of the survivors. De Soto then scuttled the ship, thinking to leave no evidence of the attack, but the prisoners managed to escape and prevent the *Morning Star* from sinking.

De Soto's crimes caught up with him after the *Black Joke* was wrecked off Spain. When he and his men arrived in Gibraltar they were recognized and taken for trial in Cadiz, where de Soto was hanged. His head was then stuck on a spike as a warning to others.

PEDRO GIBERT

SOUTH AMERICAN PRIVATEER

BORN: Details unknown, c.1800. **PIRATE CAREER:** c.1832–34. **SHIP:** *Panda*. **WHERE ACTIVE:** Caribbean. **PRIZES TAKEN:** US, Spanish and other shipping. **DIED:** Hanged, Boston, USA, 11 March 1834.

Pedro Gibert was the captain responsible for the last recorded act of piracy in the Atlantic. This took place on 20 September 1832, when his schooner *Panda* seized the US brig *Mexican*, carrying a shipment of silver from Salem to Rio de Janeiro. 'Don' Gibert, a former privateer for the Colombian government based off Florida's east coast, acted with brutality towards the crew, beating and torturing them before locking them below decks while his men searched for plunder. He set fire to the vessel and sailed away, but the crew escaped and managed to sail to safety.

The end for Gibert came after the *Panda* sank in battle with *HMS Curlew* off west Africa. The British extradited him to Boston, where this last of the Atlantic pirates was tried and hanged with three of his crew. When asked by his men what he intended for the captured crew of the Mexican, Gibert is said to have replied: 'Dead cats don't mew. You know what to do.'

SHAP'N'TZAI

CHINESE PIRATE

Born: Details unknown. **Pirate career:** 1843–49. **Where active:** South China Sea. **Prizes taken:** Chinese, British, American shipping. **Died:** Details unknown.

Shap'n'tzai was the last of the notorious pirate chiefs who terrorized the South China Sea in the early nineteenth century. Assisted by his lieutenant Chui Apoo, he made a fortune from the illegal opium trade and built up a fleet numbering hundreds of junks. His reign of terror in the seas around Hong Kong came to an end in October 1849 after his fleet was cornered by British steam-powered warships in the Hongha River in Vietnam. Shap'n'tzai thought he was safe from attack, but was unable to return fire because of the strong currents and watched helplessly as his fleet was destroyed. He escaped and made peace with the authorities. Apoo was betrayed and captured.

Survivors killed

Many of the survivors of the destruction of Shap'n'tzai's fleet were murdered by Vietnamese peasants as they struggled ashore.

PART THREE

The ships

The success of a pirate crew depended largely upon their vessel and also the skill with which it was employed. The capabilities of pirate ships over the centuries differed widely, according not only to their size and design but also to the extent to which they outclassed other seagoing vessels.

Spanish galleon

Spanish galleons carrying gold and treasures from the New World were a tempting prize for the many pirates who operated throughout the Caribbean in the Golden Age of piracy.

PIRATE SHIPS

The design of pirate ships varied greatly over the centuries, but certain general principles held true for them all. An ideal pirate ship of any age needed to be fast, well-armed and capable of holding a substantial cargo of booty. Ideally, there also had to be space to accommodate a large crew to man the sails (or oars),

and to conduct coastal raids or to board and then man any prizes taken. To make escape easier, it was also preferable for a pirate ship to have a shallow draught, enabling it to cross waters where other vessels would run aground.

With a fore-and-aft rig, and a sleek, narrow hull, the schooner was capable of speeds up to 12 knots (22 kmh), and was an American variation on the sloop.

SPEED AND STRENGTH

Speed was perhaps the most important essential, as the need to outrun both prey and pursuers was paramount. This was reflected in the sleek design of vessels from the bireme (see opposite) of the ancient Mediterranean and the galleys (see page 108) and caravels of the corsairs to the sloops (see page 119) and schooners (see page 120) of the pirates of the Caribbean as well as the junks of the South China Sea.

Alternatively, opting for a vessel that was so large and powerful that few would dare to confront it was the key to success for pirates of the Golden Age such as Blackbeard and Bartholomew Roberts, with their three-masted square-riggers (see page 123).

CONVERSIONS

Unless, rarely, pirates had the means to purchase their own vessel, they typically adapted captured merchantmen or warships for their own purposes. Such conversions might involve the removal of the forecastles or sterncastles to make room for boarding parties or extra armaments, changing the arrangement of the sails for greater speed and manoeuvrability, or even cutting gunports into the hull for extra cannon.

BIREME

This form of galley was favoured by the pirates of the ancient world. Relatively small, sleek and open-decked, it was powered by 10 to 25 oars arranged in two tiers on each side. Later versions (called hemiolas) also had a large square sail. Biremes were fast and could easily outpace most prey. With their relatively shallow draught, they could go where larger warships could not, allowing pirates to evade pursuers, but had sufficient storage space for captured cargoes.

SUPREMACY OF THE BIREME

The bireme was preferred by pirates to the lumbering triremes, with three banks of oars, and quinqueremes, with five banks of oars, of the navies that sought to protect seagoing merchant vessels. There were some variations upon the basic bireme. The Illyrian pirates sailed in open-decked biremes called lembos, while the Cilician pirates roamed the seas in liburnians, which were larger and more like naval warships. It was only when the Roman navy under Pompey the Great was equipped with similar bireme designs that the pirates were seriously challenged. Even after their threat had been largely defused the Romans kept a fleet of biremes to suppress piracy in the future.

VIKING LONGBOAT

Although the Viking longboat shared some of the basic features of the craft that plied the seas in the ancient world, with combined oar and sail power, it had various significant improvements. Carrying around 50 men and with ample space for plunder, the typical Viking longboat (several examples of which have been found and preserved) was long and sleek, with a raised point at each end. The addition of a keel gave extra strength to the longboat and

Viking longboats, their prows decorated with fearsome dragon's heads, were shallow and fast.

A Viking burial

When Viking chiefs died their bodies were often laid in their ship and either buried or burned.

(together with improved knowledge of navigation) meant that for the first time seafarers were no longer obliged to keep to shorelines for safety but could venture out into the open sea.

FAST AND MANOEUVRABLE

Longboats were fast and easy to manoeuvre, and with their relatively shallow draughts they could be beached at virtually any point along a shoreline, thereby adding to the element of surprise. They were equipped with a single row of oars on each side and with a large square sail for additional speed at sea.

Decoration

Viking crews are traditionally supposed to have hung their shields in a row along the sides of the craft, adding to its warlike appearance. They also decorated the prow and other parts of the vessel with carvings or gold and silver patterns. The prow ended in a curl, or else was ornamented with a carved dragon's head or similar figurehead.

CORSAIR GALLEY

While other pirates in the fifteenth century and later on opted for sail-powered ships, the corsairs of the Mediterranean favoured oared galleys that were directly descended from classical designs. These long, sleek vessels could be driven through the waves at an impressive rate of nine knots (16 kmph/10 mph). The power was provided by 20 to 30 oars, each manned by between three and six oarsmen, who were often captives from other ships or slaves. The galleys were also equipped with one or two sail-carrying masts that could provide added speed.

GALLEY DESIGN

The galleys often carried a battery of cannons, which usually fired forwards only and were not the preferred means of attack as artillery fire would damage a potential prize. Swivel guns mounted elsewhere on the vessel gave extra firepower at close range. With their superior speed and manoeuvrability the corsair galleys outmatched the lumbering sailing ships they were pitted against, despite their inferior artillery.

There were several variations in the basic galley design, with different shapes of hull and different sizes of gun platform. Most corsairs sailed in a smaller, single-masted version of the galley called a galiot, which boasted even greater speed. On these ships the corsairs took on the duties of oarsmen, too.

Galley slaves

The oarsmen lived a terrible life, suffering from the harsh demands of their work and the cruelty of their overseers; many died at their seat in the hull. They were not expected to fight, however, as this was left to well-armed Janissaries.

Galleys in their various forms were still in use by Barbary corsairs into the early nineteenth century.

GALLEON

The galleon opened up the world's oceans to the seafarers of the European nations. Until the sixteenth century, most seafaring was confined to inland seas like the Mediterranean or to the coastlines of Europe and beyond. Few ships ventured far into the open sea.

The galleon was developed initially out of the earlier, smaller carrack by Spanish shipwrights, who needed to build vessels that were capable of bringing the treasure of the New World home across the Atlantic. Spanish treasure galleons were typically large, top-heavy, ungainly ships, which were designed to carry huge cargoes and large numbers of soldiers and passengers, although some were smaller and more manoeuvrable. The very biggest were reserved for the Spanish navy, while those that guarded the treasure fleets were generally smaller in size, though still made cumbersome by their large holds.

SPANISH TREASURE GALLEON

Length: 38 metres (125 feet). **Width:** 12 metres (35 feet). **Weight:** 180–450 tonnes (200–500 tons). **Displacement:** Up to 1100 tonnes (1200 tons). **Speed:** 4–8 knots. **Firepower:** Up to 60 guns. **Crew:** Up to 200.

The Spanish treasure fleet that sailed annually to Spain between 1530 and 1735 was the ultimate prize for any pirate but in the end it was not piratical activity that ended the treasure fleet system but a fall in the price of silver after Spain over-supplied the market.

The Spanish treasure galleons that were such a lure to privateers and pirates in the sixteenth and the seventeenth centuries were large, three-masted, square-rigged ships which were capable of carrying huge quantities of gold and other cargo.

The Spanish galleons differed from their English and Dutch equivalents not only in their size, but also in their superstructure. They had tall forecastles and large, towering sterncastles, which made them top-heavy. Most of the bulk of the ship was above the waterline, as the ships had to load and unload in shallow water. To maintain stability, the hulls tapered sharply inwards to a fairly narrow top deck.

A formidable foe

The treasure galleons were relatively slow and unmanoeuvrable. To counter this they carried many guns and large numbers of soldiers to board any attacking vessel that came too close. Treasure ships

Evolution of the pirate galleon

The English galleons were in their own turn copied by Dutch shipwrights and it was from these that the typical pirate galleon of the seventeenth century evolved.

often sailed in vast convoys and it was a brave captain who risked their massed firepower.

THE ENGLISH GALLEON

English shipwrights quickly copied the Spanish designs, but they made their version of the galleon smaller, sleeker and faster, thereby gaining a distinct advantage in battle with the lumbering treasure ships of the Spanish Main. English galleons also had better guns and they could maintain a higher rate of fire than their clumsier Spanish counterparts. Such galleons would not only become the backbone of the Royal Navy, but they would also carry Sir Francis Drake and other privateers to distant parts of the globe in search of plunder.

GOLDEN HIND

LENGTH: 23 metres (75 feet). **WIDTH:** 6 metres (20 feet). **WEIGHT:** 102 tonnes (100 tons). **DISPLACEMENT:** 109 tonnes (120 tons). **FIREPOWER:** 18 guns. **CREW:** 164. **CAPTAIN:** Sir Francis Drake.

The *Golden Hind* was the small English-built galleon in which Sir Francis Drake sailed to the Spanish Main in 1579. It was typical of English galleons of the period, being small, streamlined and fast. It was also

Patronage

Sir Francis Drake called his ship the *Golden Hind* in tribute to his patron, Christopher Hatton, whose heraldic shield included a hind (roe deer).

well-armed, with 18 guns of various sizes, and more manoeuvrable than the Spanish treasure galleons.

English galleons like the *Golden Hind* had a relatively low superstructure, with a modest quarterdeck, thus maintaining a sleek, seaworthy shape. They also weighed much less than the treasure ships. The fact that the English guns had better carriages than the Spanish versions meant that English gunners (who were also better trained) could keep up a faster rate of fire. In battle Drake, like other English captains, kept his distance from his prey, relying upon his ship's superior firepower to secure victory.

Loss and rebuilding

Having sailed right round the globe, the *Golden Hind* remained on public display in London until its timbers finally rotted away. In 1973, a full-scale replica of the vessel was built to recreate Drake's famous circumnavigation before being installed as a museum on the River Thames.

BRIGANTINES, SLOOPS AND SNOWS

Pirates of the Golden Age of piracy that spanned the end of the seventeenth century and the start of the eighteenth century set to sea in whatever vessels they could lay their hands on. Very few of them had the money to purchase a substantial craft but had instead to steal or seize whatever they could find, often through mutiny.

Typically starting with a small sailboat, such as a pinnace or flyboat, they would then hope to trade up by capturing ever larger vessels until ultimately they had command of a three-masted square-rigger as large as any warship afloat.

TYPES OF CRAFT

There were many different types of smaller vessel for pirates to choose from. The most modest of vessels was the simple longboat (a rowing boat that could carry half a dozen or more men).

From this a pirate might progress to a fast sloop, the vessel with which most pirate attacks were made, or to a schooner, an American version of the sloop

favoured by American privateers, or a similar two-masted ketch. Such swift, versatile craft were ideal for piracy in inshore waters. Technically speaking, these smaller vessels were not 'ships', a term reserved in the eighteenth century solely for three-masted vessels with square-rigged sails.

A two-masted naval brigantine of the type that was often sailed by pirates.

Another class of vessel was the brigantine, a two-masted vessel combining speed with firepower. Variants of the brigantine, distinguished by different arrangements of the sails, included the brig and the snow. From such craft it was but a small step to the three-masted flagship of such captains as Blackbeard and Bartholomew Roberts.

PINNACE

The pinnace was a relatively small sailing vessel of a type much used in the late seventeenth century by such buccaneers as Sir Henry Morgan. Originally the term referred to a simple oared longboat, rarely more than 11 metres (35 feet) in length, with a single mast on which would be set a single triangular sail. Some pirates kept such pinnaces in tow behind larger vessels for use in coastal raids.

Buccaneer craft

The term pinnace also came to be applied to a somewhat larger class of vessel, which could have a displacement in the water of as much as 200 tonnes (197 tons). These larger pinnaces had between one and three masts and the rig of the sails varied from one vessel to another.

Some were large enough to carry as many as 20 guns. These were the pinnaces which Sir Henry Morgan and other buccaneers captained in the Caribbean.
Some equivalents of the smaller variety of pinnace in

other countries included the Dutch *pingue* and the Spanish *barca longa*.

SLOOP

The sloop was a fast single-masted vessel with a fore-and-aft rigged mainsail and single foresail, although the arrangement of the masts and sails varied. The sloop, which developed out of the earlier pinnace, became the commonest and most successful form of pirate vessel.

Usefulness to pirates

Most pirates began their careers in sloops before progressing onto larger vessels, finding them the ideal craft for chasing down their prey and then making a swift escape. Even those who went on to command three-masted square-riggers often still kept sloops to conduct combined operations. A typical sloop might have a deck of some 20 metres

Sloops in the Golden Age

During the Golden Age of piracy in the early eighteenth century, more than half of all the pirate attacks in American waters were made with sloops.

(65 feet) in length and a beam (width) of 6 metres (20 feet). There was usually no quarter-deck, but an upward slope in the deck towards the stern.

Specifications

Most sloops could carry around a dozen cannons and were manned by a crew of about 75 men. In addition to sails, some sloops could also be propelled by oars when there was no wind. A variant was the single-masted cutter.

SCHOONER

The two-masted schooner emerged as an American variant of the sloop and it is often presented as a typical pirate craft in fiction, although in reality such vessels were very rarely seen in pirate hands in the Golden Age of piracy in the early eighteenth century. No records exist of pirates using schooners in attacks before 1717, and only five per cent of pirate attacks in North American waters between 1717 and 1730 involved the use of schooners.

Sleek and fast

With a fore-and-aft rig, sometimes with an extra square topsail for added speed, and a sleek, narrow hull, the schooner was capable of catching up with all but the fastest of prey and had a top speed of

The age of the schooner

The schooner really came into its own towards the end of the eighteenth century, when purpose-built heavy schooners which were constructed in Baltimore became the favourite craft of American navy captains as well as privateers such as John Paul Jones.

12 knots (22 kmh). Like the sloop, the schooner had a relatively shallow draught and could thus go where many larger vessels could not, making escape and concealment much easier.

A typical schooner carried around eight cannons and was manned by a crew of 75. The displacement could be as much as 91 tonnes (100 tons).

BRIGANTINE

The brigantine was a relatively large two-masted craft, which could be distinguished by the square-rigged sails on its foremast and combined triangular 'gaff-rigged' sail and square sail on its mainmast. It might be up to 25 metres (80 feet) in length, could carry a dozen cannons and was manned by a crew of around 100 men.

Uses of the brigantine

Brigantines were widely used for coastal trading on the American side of the Atlantic but they were not the favourite vessel of most pirates as they could not match the high proportion of sail to displacement that was offered by the more common sloop.

Nonetheless, a number of famous pirates sailed in brigantines in the Caribbean and elsewhere in the early eighteenth century, among them Charles Vane, in the brigantine *Ranger*.

Variations

Some variations on the brigantine were as follows:

- The brig, which carried square-rigged lower sails on both masts as well as a triangular fore-and-aft sail.

- The square-rigged snow, which had a secondary mast with a fore-and-aft sail which was positioned immediately behind the mainmast.

Derivation of the name

The brigantine possibly owed its name to its original association with the buccaneers (or 'brigands') of the seventeenth century.

PIRATE FLAGSHIP

Only a few pirate captains, such as Blackbeard and Captain Kidd, enjoyed the luxury of commanding a large, square-rigged warship the equal of any other ship on the high seas.

They had three masts, carrying square-rigged sails, and were impressively armed with up to 40 guns. Large crews were needed to sail them and, in some cases, to man the oars when there was no wind. A broadside from Blackbeard's *Queen Anne's Revenge* required 60 men working as gun crews, with four to six men at each cannon. Such flagships were slower than smaller vessels such as sloops, but they were usually more seaworthy and thus preferred by pirates who aimed to pursue their trade on the open ocean, rather than closer inshore.

Famous flagships

The most famous pirate flagships during the Golden Age of piracy were Blackbeard's *Queen Anne's Revenge*, Captain Kidd's *Adventure Galley*, Samuel Bellamy's *Whydah* and Bartholomew Roberts' *Royal Fortune*.

QUEEN ANNE'S REVENGE

LENGTH: 33 metres (110 feet). **WIDTH:** 7 metres (24 feet). **WEIGHT:** 203–305 tonnes (200–300 tons). **FIREPOWER:** 40 guns. **CREW:** up to 300. **CAPTAIN:** Blackbeard.

Blackbeard chose for his flagship a 14-gun three-masted French slaving ship called *La Concorde*, which fell into his hands off St Vincent in 1717. He renamed it and fitted it out for attacking other vessels, with extra gunports cut into the hull for 26 more guns. It was manned by 300 men and became one of the most powerful ships in the Caribbean. Its very appearance struck terror into the crews of its victims, who often surrendered without firing a shot. In 1718, it ran aground on a sandbar and was abandoned, along with much of the crew. It is probable that Blackbeard did this deliberately as a ruse to strand many of his men so there were fewer with whom to share his booty.

Identifying a wreck

In 1996, a wreck that was thought to be that of the *Queen Anne's Revenge* was discovered at Beaufort Inlet, North Carolina, in the USA. It has since been extensively examined by marine archaeologists (see page 243).

ADVENTURE GALLEY

Length: 35 metres (114 feet). **Width:** 9 metres (28 feet). **Weight:** 292 tonnes (287 tons). **Firepower:** 34 guns. **Crew:** 70–152. **Captain:** William Kidd.

Built in Deptford, London, in 1695, the *Adventure Galley* was an unusual combination of three-masted galleon and oar-powered galley. The ship was designed as a privateer but proved equally ideal as a pirate vessel, with a daunting array of cannons and a top speed of 14 knots. When there was insufficient wind to fill the square-rigged sails, with their 3,200 square yards of canvas, the vessel could be propelled by means of 46 oars (or 'sweeps').

Privateer to pirate ship

The ship was paid for by a group of wealthy London backers so Kidd could sail to the Indian Ocean with a privateering commission to subdue pirates in the region. Once there, however, he turned pirate himself, and seized many prizes, the most notable being the *Quedah Merchant*, which carried a hugely valuable cargo. In 1698, while on his way home, Kidd found that the ship's hull was rotten and abandoned it in a bay in north-eastern Madagascar. He transferred his booty to the captured *Quedah Merchant,* then set fire to the *Adventure Galley* and sailed for New York.

CHINESE JUNK

To European eyes, the wooden sailing vessels of the Chinese pirates appeared primitive and strange, but they were soon revealed to be manoeuvrable and formidably armed. Chinese pirate junks dominated the seas off China for hundreds of years, regardless of repeated attempts to sweep them from the oceans by the Chinese and other navies.

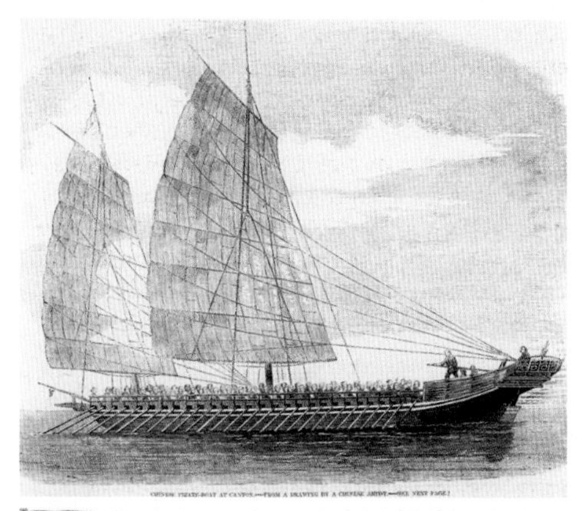

The distinctive shape and sails of the Chinese junk makes it an unmistakable ship.

Like pirates elsewhere, Chinese pirates generally converted captured merchant vessels, cutting gunports in the hull for cannons and mounting swivel guns on the rails. The junks varied in size, some measuring up to 30 metres (100 feet) in length. The largest had three or more masts and between 10 and 15 guns. The square sails were made from bamboo matting.

PIRATE FLEETS

Chinese pirates often sailed in large squadrons, each with its own flag, and terrorized shipping and coastal settlements. They usually attacked by boarding, as gunfire might endanger the safety of their prize. Crews, who might number as many as 200 men, often sailed with their families on board, the captain living in his own cabin in the stern and the rest in the hold. It was only with the arrival of large forces of steam-powered European naval vessels that the junks, whose design had remained essentially unchanged for centuries, finally found themselves outgunned and outclassed.

The first junk

Chinese legend claims that the first junk was designed by Emperor Fu His, who was himself the son of a sea nymph.

FAMOUS PIRATE SHIPS

The names of some pirate ships have long since passed into folklore. Names of captured vessels were often changed on being seized by pirates, largely in order to conceal their real identity.

For the same reason, the name of a pirate ship might be changed from time to time at the whim of its captain. As pirates traded up from a smaller vessel to a larger one, they would often simply keep the same name and apply it to the new ship.

Ship	Captain
Adventure Galley	William Kidd
Amity	Thomas Tew
Black Joke	Benito de Soto
Bonhomme Richard	John Paul Jones
Duke	Woodes Rogers
Fancy	Henry Every, Edward England
Flying Dragon	Edmund Condent
Golden Hind	Sir Francis Drake

Ship	Captain
Happy Delivery	George Lowther
Jesus of Lubeck	Sir John Hawkins
La Confiance	Robert Surcouf
Le Roi David	Jean Bart
Liberty	Thomas Tew
Oxford	Sir Henry Morgan
Pearl	Edward England
Queen Anne's Revenge	Edward Teach (Blackbeard)
Ranger	Charles Vane
Revenge	John Gow, Stede Bonnet, John Phillips, Edward Teach (Blackbeard)
Rising Sun	William Moody
Royal Fortune	Bartholomew Roberts
Saint James	Howell Davis
Trader	Charlotte de Berry
Trinidad	Bartholomew Sharp
Whydah	Samuel Bellamy

PARTS OF A PIRATE SHIP

In order to sail their ships efficiently pirates had to be familiar with every part of their vessels, all of which had their own names.

1 anchor
2 armoury
3 ballast
4 bow
5 bowsprit
6 captain's cabin
7 colour

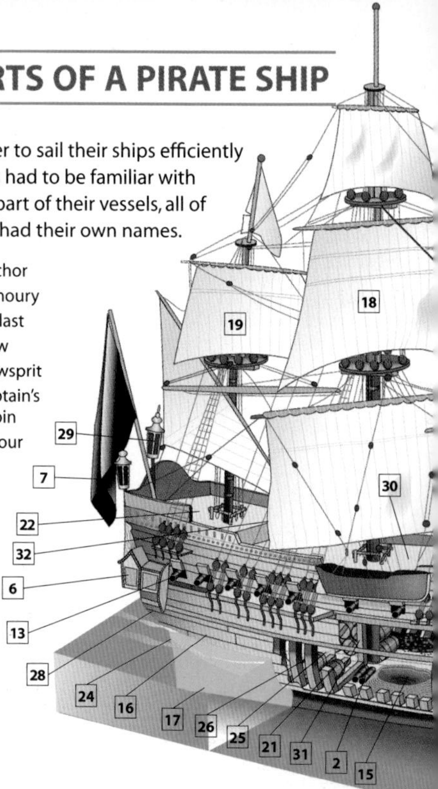

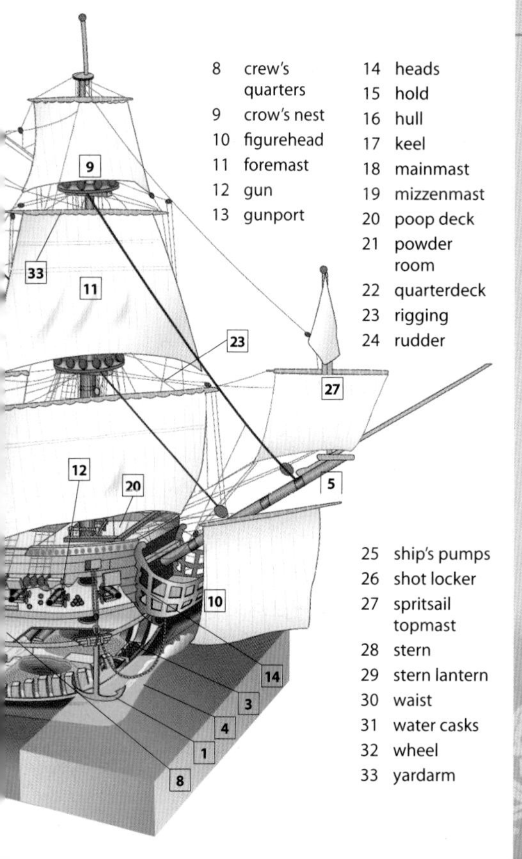
8 crew's quarters
9 crow's nest
10 figurehead
11 foremast
12 gun
13 gunport
14 heads
15 hold
16 hull
17 keel
18 mainmast
19 mizzenmast
20 poop deck
21 powder room
22 quarterdeck
23 rigging
24 rudder
25 ship's pumps
26 shot locker
27 spritsail topmast
28 stern
29 stern lantern
30 waist
31 water casks
32 wheel
33 yardarm
9
33
11
23
27
12
20
5
10
14
3
4
1
8

EQUIPMENT

A pirate ship would be crammed with hundreds of items of equipment which were essential to the good running of the ship. Besides weapons and personal effects, there were many tools, above and below decks, in almost daily use. The following selection represents the range of equipment that all pirate ships would be expected to carry, as proved by what has been found in pirate wrecks over the years.

ADZ

Resembling a workman's pick, this was one of the essential tools used by pirate crews when careening (cleaning) the wooden hulls of their ships, the blade being used to chip away barnacles.

ANCHOR

The ship's anchor kept a ship in a fixed position, its points lodging in the seabed and preventing the ship being moved by wind or tide. In the case of a large ship, it could take as much as an hour to haul the anchor up before setting sail. If a pirate ship was caught at anchor by enemy vessels one option was to cut the anchor cable to enable a quick escape.

CALTROP

This comprised three spikes arranged in such a way that however the weapon fell one spike would always point upwards, making it capable of causing terrible injuries. Such caltrops, originally used against cavalry, were tossed by French corsairs onto the deck of an enemy vessel to hinder the actions of the crew, who were usually barefooted.

CAT-O'-NINE TAILS

A whip with nine strands of knotted rope was used to flog crew members who had incurred the displeasure of their captain. Traditionally it was the duty of the man to be flogged to make the whip that would be

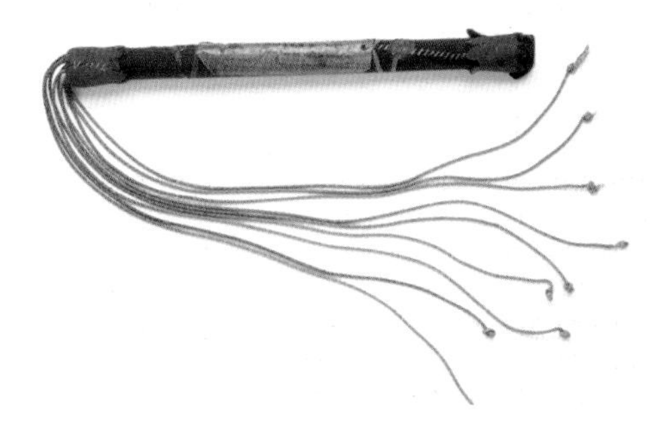

used on himself. Cat-o'-nine tails were not supposed to be used more than once, because blood soaked into the strands could spread infections.

GRAPPLING IRON

This consisted of a four-pronged metal spike with barbed points tied to a rope. It was thrown onto an enemy vessel so as to lodge in the rigging and allow the pirate crew to pull it closer, ready for boarding.

MEDICINE CHEST

Medical treatment would have been rudimentary in the extreme. Most crews included men who were diseased or risked serious injury on a daily basis and thus if a captured ship had a medicine chest it was likely to be seized. Blackbeard himself demanded medical supplies as part of his ransom when he was blockading the port of Charleston in 1718.

SPEAKING TRUMPET

Most pirate captains included in their personal equipment a speaking trumpet with which to give orders to the crew or to shout demands to nearby ships. The shape of the trumpet amplified the sound of the voice.

WEAPONS

The pirates who sailed the Mediterranean in ancient times fought with swords, spears, knives, axes, slings and bows. This did not change until the introduction of musket and cannon in the sixteenth century brought about a revolution in warfare on land and at sea.

CANNONS

Pirate ships carried as many as 40 or more cannons of different calibres. Most were mounted on wooden carriages and fired either from the open deck or

Pirate ships carried as many as 40 or more cannons of different calibres.

through gunports which were cut into the side of the hull. The elevation was altered by wedges under the rear end of the barrel.

Smaller cannons, known as minions, fired balls weighing three or four pounds. The largest cannons fired 32-pound balls, but these were rare and eight- or twelve-pounders were the preferred calibre. Cannon balls shattered timber and brought down masts and sails. When fired into the hull of a ship an iron cannon ball blasted deadly wooden splinters through the vessel, causing many injuries.

Other forms of ammunition included chain shot, in which two small iron balls were linked by chain. This form of shot was particularly effective against rigging and could be used to slow a ship down. Canister or case shot (a container of small iron balls or stones that split open on being fired) was deadly at close range.

Warning shots

Pirates were generally reluctant to use cannons against vessels they hoped to capture and then sell or take over themselves. Often pirates would fire a single cannon as a warning to a merchant ship to surrender or be blown to bits: they often gave in at once. Some pirate sloops, indeed, had no more than two cannons, mounted in the bows for such purposes.

SWIVEL GUNS

The smallest calibre of cannon was the swivel gun, usually mounted on a ship's rails. It was used chiefly to clear the decks of enemy ships prior to boarding or to repel hostile boarding parties.

PERSONAL WEAPONS

The pirates' choice of hand weapons was influenced by the likely circumstances in which they would be used. The surroundings of a crowded ship's deck, with all the obstructions of the rigging and other nautical equipment, meant that the best weapons were those that could be used in cramped conditions and with great effectiveness at close quarters.

THE CUTLASS

This remains the trademark weapon of the archetypal pirate. It was the usual weapon carried by seamen of all kinds from the fifteenth to the eighteenth century. Its short, broad blade made it ideal in hand-to-hand fighting on board ship and cutlasses were carried by captains and crew members alike. The cutlass was supposedly invented by the original buccaneers, having its origin in the long knives they used when butchering meat for their *boucan* barbecues.

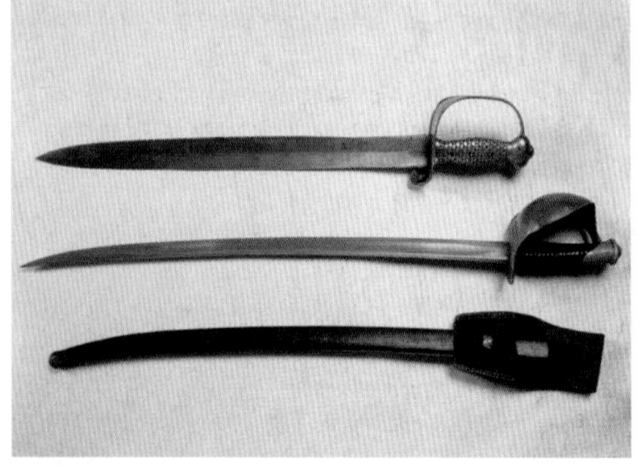

With their short, broad blades, cutlasses were the ideal weapons in hand-to-hand fighting.

MUSKETS AND PISTOLS

Muskets and pistols of the Golden Era of piracy in the early eighteenth century were relatively inaccurate, except at close range, but lethal when ships came within boarding range. The main drawback was that they took a long time to reload, so after firing the first shot they were usually employed as clubs. A variant of the musket that proved particularly effective at close quarters was the musketoon, a short-barrelled musket with a flared muzzle that was easier to wield in the cramped surroundings of a crowded deck.

Flintlock pistols were carried by most pirate captains as well as many of the crew – Blackbeard was famous for carrying no less than six. Though only effective at close range, they were commonly used by boarding parties and after being fired made a handy club.

OTHER HAND WEAPONS

Other weapons commonly used by pirates included daggers, axes and rudimentary grenades. Axes were useful to boarding parties when climbing up the high wooden wall of a ship's hull.

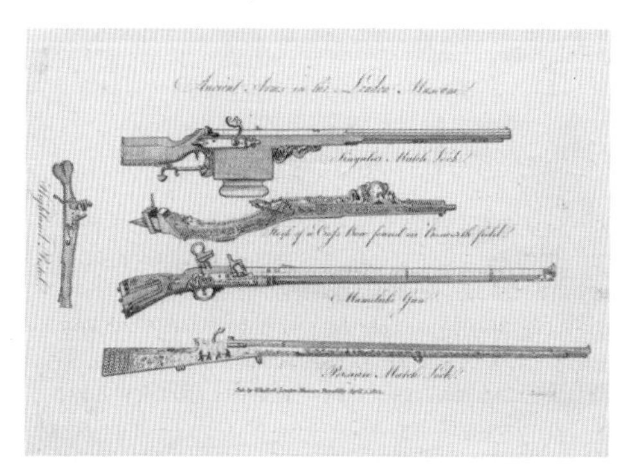

Muskets and pistols of the Golden Era of piracy were relatively inaccurate except at close range.

TACTICS

Pirates in the ancient world generally subdued their victims by drawing close to their vessels and then boarding them. If the pirate galley had a sail this was lowered to become a fighting platform from which the pirates might leap onto their prey. Speed was the best weapon in their armoury, as most merchant vessels were slower and less manoeuvrable.

In the medieval period battles at sea continued to be fought at close quarters with swords, axes and bows and arrows. The corsairs of the Mediterranean fought their battles at similarly close quarters as late as the early nineteenth century, as did the Chinese pirates of the South China Sea.

DECEPTION

In the fifteenth century the design of stronger hulls meant that cannons could be carried and boarding was no longer the only option. Better guns and gunnery training gave early English privateers like Sir Francis Drake an advantage over Spanish treasure ships, despite the fact they generally had fewer guns. The usual tactic was to bombard the prey from a distance, allowing the Spanish no chance to board.

Pirates were often reluctant to use their guns on a potential victim, as this could result in the loss of the ship and its cargo; subtler means were preferred.

Surrendering without a fight

A favourite ploy was to intimidate a ship into surrendering, usually by hoisting a pirate flag and firing a warning shot. Fear of pirates and what happened to those who resisted them meant that victims often surrendered without a fight. On other occasions, pirates would fly false colours and draw close to their prey before revealing their true identity, or steal silently up to a ship in small boats and then swarm quietly aboard to take the crew by surprise.

BEING IN THE RIGHT PLACE

The most successful pirates knew where to find the richest prizes. In the Caribbean, pirates lay in wait off eastern America for Spanish treasure galleons sailing up the coast seeking favourable winds to blow them across the Atlantic. Others established their base in Madagascar, where they would be in easy reach of the trade routes to India and the Far East.

Overleaf (pages 142–143): Sometimes pirates would fly false colours and draw close to their prey before revealing their true identity.

NAVIGATIONAL TOOLS

The first pirates, like other early seafarers, had only a limited understanding of how to find their way around the seas. Because of this they rarely strayed out of sight of land. Later, however, advances in the understanding of astronomy and navigation, coupled with the development of new navigational tools, meant that pirates could venture into the open ocean in search of their prey.

SEA ARTISTS

The skills of the navigators, or 'sea artists', remained relatively basic until the sixteenth century, when major improvements in maps and equipment were made and voyages across oceans became possible. As well as better maps, navigators now had a range of tools available to them: telescopes, compasses, dividers (for transferring measurements between maps and scales) and increasingly sophisticated instruments for calculating longitude and latitude (the ship's position on a map in terms of east or west and north or south respectively).

Some pirate navigators were better equipped than others, depending largely upon what they managed to steal from the vessels they seized, so instinct, luck and local knowledge also played a significant role. Navigating in the Caribbean, with its many islands, was relatively easy, but finding one's way in open oceans was more of a challenge.

COMPASS

The most important piece of navigational equipment was a compass, which located due north and could

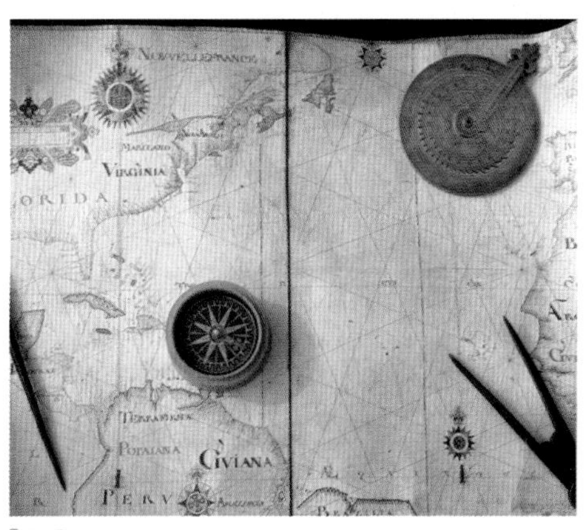

Improvements in navigational tools opened up the New World to seafarers of all kinds, including pirates.

be used to estimate longitude. Before the invention of the compass the only way to judge direction was by the position of the sun and the stars. Pirates could make crude compasses themselves by passing a piece of magnetic lodestone over a needle so that it became sensitive to the earth's magnetic field.

TELESCOPE

Invented in the sixteenth century, the telescope (or 'bring 'em near') was very useful to both pirate navigators and captains. While captains used them to search for and identify other ships, navigators could employ telescopes to examine distant landmarks or to detect clues such as cloud formations indicating where land lay.

BACKSTAFF

Early pirates of the Spanish Main originally used a handheld instrument called a cross-staff to calculate latitude (their position in terms of north or south) based on the position of the sun in the sky. However, the cross-staff was difficult to use as it required the navigator to stare directly at the sun, so in 1595 a better tool, called the backstaff, was invented for the same purpose by English navigator John Davis, who had sailed with the privateer Thomas Cavendish in

1591. To use this instrument a navigator stood with his back to the sun and measured its shadow instead.

TOOLS OF THE TRADE

Some other useful mechanical aids to navigation occasionally fell into the hands of pirate navigators. These included globes of the world showing (more or less accurately) the shape and position of the major land masses and oceans, astrolabes (an instrument for measuring the altitude of stars and planets), sundials and wind vanes (which indicated the direction of the wind).

Sometimes several of these tools were combined as one instrument, called an astronomical compendium, although these were expensive and thus rare.

BOOKS

Navigators tended to be relatively well-educated men who could read, at least after a fashion. Several books were published which offered practical help with navigation. The most important of these was probably John Davis' *The Seaman's Secrets*, which incorporated a *volvelle* (a circular diagram by which tides could be worked out from the phase of the moon).

MAPS

Possession of a good map was crucial to the success of many a piratical voyage. Most pirate maps were less than accurate and thus captains relied as much upon local knowledge and keeping a good watch as upon their charts in order to avoid shipwreck. A good map (or 'waggoner' as they were sometimes called) was thus of great value to pirates.

CAPTURED CHARTS

The Spanish had charted much of the New World coastline early in the sixteenth century, and therefore the charts that were taken from captured Spanish ships were always greatly prized.

Sharp's charts

The most famous of these to fall into the hands of pirates was the book of maps seized from a Spanish ship by Bartholomew Sharp in 1681. The Spanish crew attempted to throw the charts overboard but Sharp prevented this and took them for his own, upon which the Spanish captives are said to have burst into tears. The maps proved so detailed that they were taken back to England and there copied for use by other English captains.

FLAGS

The 'Jolly Roger', depicting a skull and crossbones, is universally recognized today as the archetypal flag flown by pirates in the seventeenth and eighteenth centuries. However, in reality, there were many different versions of the Jolly Roger, each of which was unique to a particular captain.

The designs were deliberately bloodthirsty, with depictions of skulls, skeletons, cutlasses, bleeding hearts and hour glasses, so that potential victims would be left in no doubt of the threat facing them if they did not surrender immediately. The inspiration for the designs is thought to have come from gravestones bearing similar emblems of death.

False colours

Pirates often waited until the last moment before revealing their flag, sailing close to their victim under the colours of a friendly nation so that the crew's suspicions would not be aroused. When the flag was finally unfurled, the shock was calculated to persuade the victim to give in without a fight.

CHINESE FLAGS

Chinese pirates also flew distinctive banners. The colour of each flag was unique to a particular pirate squadron, and it might bear elaborate designs depicting the gods or symbols of good luck.

No quarter

The Jolly Roger is thought to have got its name from the French *joli rouge* ('pretty red'), a reference to the red flag (called the 'Red Jack') that was flown by early English privateers. When pirates flew a plain red flag this would be interpreted as a warning that no mercy would be shown to any defeated enemy.

INDIVIDUAL FLAGS

Records have survived of some of the flags that were flown by celebrated pirates. One flag that requires some explanation is one of those flown by the pirate Bartholomew Roberts: the initials ABH and AMH, positioned below two skulls crushed under a pirate's feet, stood for 'A Barbadian's Head' and 'A Martinican's Head' and were a warning to the inhabitants of those two Caribbean islands, who had incurred his wrath.

Bartholomew Roberts

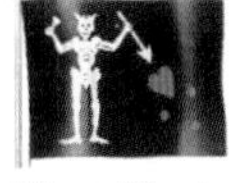

Edward Teach (Blackbeard)

Christopher Moody

Edward England

Edward Low

Henry Every

Jack Rackham

Thomas Tew

Walter Kennedy

PART FOUR

The pirate life

In the popular imagination pirates lived colourful lives of high adventure, with the additional possibility of acquiring great wealth and a lasting position in maritime lore. In reality, the lives of most pirates were far less attractive and liable to end prematurely in battle, of disease or at the end of a hangman's rope.

Pirate booty

The spoils of captured ships were shared out among the pirate crews. Spanish treasure galleons, loaded with gold, were a particularly tempting prize for Caribbean buccaneers.

LURE OF THE JOLLY ROGER

The life of a pirate was hard and often short, but it also offered the prospect of freedom from authority and, for the lucky few, immense wealth and the opportunity of social advancement. Such rewards made the risk of death in battle or by execution an acceptable gamble for many.

The life of a captive in a pirate galley in the classical era was unlikely to be very long, and adventurers in the days of the buccaneers and during the Golden Age of piracy in the early eighteenth century had a similarly short life expectancy. Even infamous pirates, such as Captain Kidd and Blackbeard, had active careers lasting only a year or two.

FREEDOM AND RICHES

Hard though it was, the life of a pirate could look very attractive when it was compared to that of a penniless beggar on shore or to that of a seaman in the Royal Navy, who was subject to harsher discipline and stood no chance of earning a fortune.

The lure of great riches, combined with a life of comparative freedom from authority, persuaded many generations of seafarers to chance their luck as pirates. Some were privateers who had grown used to plundering enemy ships in wartime and who had no intention of giving it up now peace had been restored. Others were driven to piracy by desperation: the end of the seventeenth century found many English seamen out of work with the end of the war against France, and life as a pirate offered the chance to employ their skills for their own benefit.

Spoils from pirate raids were divided equally among the crew or shared out according to status.

THE PIRATE CREW

A typical pirate crew of the Golden Age of piracy in the early eighteenth century was drawn from a range of backgrounds, nationalities and races. Most pirates were experienced seamen in their twenties and the members of the crews of captured vessels, persuaded to switch allegiances or forced to join their captors. Most of the others were former privateers, deserters from navy or merchant vessels or mutineers, of whom many had been originally 'pressed' (forcibly kidnapped) into going to sea against their will.

NATIONALITIES

Caribbean pirates of the Golden Age were mostly English, American or from the West Indian colonies. However, there were also other nationalities, whose representatives included blacks as well as whites. Bartholomew Roberts' crew, who were captured in 1721, comprised 187 whites and 75 blacks, of whom many were runaway slaves or volunteers from captured slave ships. Few blacks joined as experienced seamen and probably sailed (initially at least) as servants in pirate crews. Whatever their nationality, most crew members were ill-educated and came from very modest backgrounds.

The act of just hoisting a pirate flag was often enough to make a ship surrender.

RANKS

Leadership was usually shared between two elected officers. The captain exercised control in battle, while his second-in-command, the quartermaster, supervised the running of the ship at other times and presided over the division of booty. Other officers were appointed by the captain, sometimes aided by his quartermaster. These might include the master, who undertook the duties of navigator, the boatswain (or bo'sun), who maintained the ship and its equipment, and the gunner, who looked after the cannons.

WOMEN PIRATES

Pirate crews were essentially male societies and thus women were banned from most pirate ships. Any woman who fell into the hands of pirates was likely to be maltreated. There were, however, a few exceptional women who fought as equals alongside male pirates, becoming part of pirate lore.

BONNY AND READ

The most famous women pirates of all were Anne Bonny and Mary Read (see page 85), who sailed with 'Calico Jack' Rackham during the Golden Age of piracy in the early eighteenth century.

Bonny and Read were examples of pirates who ran away to sea to escape a dreary life ashore. Anne Bonny ran away with Rackham after meeting him on New Providence, while Read was already serving in

Dressed as men

Both Bonny and Read had dressed as males since childhood, Read at the wishes of her mother and Bonny to satisfy the requirements of a will.

Chinese pirate queens

Chinese pirates often sailed with their wives and families, but some women were pirates in their own right. The most powerful of all was the notorious pirate queen Cheng I Sao, who had 80,000 pirates under her control.

male disguise as an ordinary seaman when she was taken prisoner by Rackham's pirates and persuaded to join them. They proved formidable fighters, and when challenged by a government sloop Bonny and Read alone resisted.

OTHER WOMEN PIRATES

A female pirate captain named Alwilda, from southern Sweden, terrorized the seas before the advent of the Vikings. She is reputed to have gone to sea with an all-female crew to avoid marriage to the Danish prince Alf but eventually retired in order to marry him after all. Other notable female pirates included Irishwoman Grace O'Malley in the mid-sixteenth century and, in the same century, the English pirate Charlotte de Berry, who followed her husband into the navy, led a mutiny and became a pirate captain (though she may be purely legendary).

PIRATE DRESS

Pirates through the centuries dressed much like other seamen. They might, however, steal clothing from the vessels they captured, if necessary stripping any garments they took a fancy to from their prisoners.

THE CREW

During the Golden Age of piracy in the early eighteenth century pirates typically wore short blue jackets ('fearnoughts'), as worn in the Royal Navy, often over a red or blue waistcoat and a plain or blue-and-white checked linen shirt. They preferred canvas or calico trousers to knee breeches, or else wore 'petticoat breeches' cut short above the ankle. These were fastened with bone buttons and might be lightly coated with tar to make them waterproof.

Earrings

Pirates are often depicted wearing earrings. It was a popular belief in times past that wearing gold or silver in the earlobe improved eyesight or protected the wearer from drowning.

Crew members often went barefoot on board ship but typically donned woollen stockings and shoes on shore. Other items of clothing might include a neck scarf, a brightly-coloured sash and a crossbelt to hold a cutlass. On their heads they might wear a tricorn hat or else a knotted scarf or woollen cap.

THE CAPTAIN

Many pirate captains liked to ape the appearance of a gentleman, priding themselves on their fine clothes, complete with plumed tricorn hats, embroidered waistcoats and lacy cuffs. The notorious Bartholomew Roberts was well-known for his flamboyant dress, which included a crimson waistcoat and breeches, a red feather in his hat and a gold chain bearing a diamond cross round his neck. Stede Bonnet also had a reputation as a dandy, wearing fine clothes and a gentleman's powdered wig. When Blackbeard saw him he burst into laughter and then assumed command of his vessel.

Tricorn hats

The stereotypical tricorn hat of the pirate captain was the standard headwear of gentlemen generally, but it lacked the skull and crossbones emblem of popular folklore; this was an embellishment invented by the writer J.M. Barrie in *Peter Pan*.

LIFE AT SEA

Although Hollywood movies make it seem glamorous and exciting, life on board a pirate ship was often miserable in the extreme, with long periods of hardship and boredom interrupted by short periods of violence and danger. Ships were often overcrowded and pirate captains could be cruel and unpredictable.

As well as fighting, crew members had to play their part in running the ship, adjusting sails, manning cannons, making repairs, working the pumps and performing the many tasks that were necessary to keep a vessel afloat. When not working they typically whiled away the time playing cards or dice.

Living conditions could be very grim. The officers might have relatively luxurious quarters towards the stern, but other crew members had to live in cramped spaces below decks that were foul-smelling, damp and uncomfortable.

FOOD AND DRINK

The pirate diet was less than appetizing. Fresh food quickly went rotten, leaving the crew to rely upon fish caught with lines, boiled sea turtles (kept alive

'Hardtack' biscuits were a long-lasting staple food for many pirates.

on board until needed) and tough biscuits called 'hardtack'. These soon became infested with maggots – called weevils – so many pirates ate in the dark where they could not see their food. Meat would be cooked in a cauldron, but because of the obvious risk of fire this could only be used when the sea was calm.

Water and alcohol

Water soon became undrinkable, so the crew drank beer, rum or grog (rum and water), or even rum and gunpowder, though some captains forbade alcohol on board ship (some crews were captured while drunk).

The risk of starving or dying of thirst was ever-present. Pirates marooned or shipwrecked on desert islands faced a slow, agonising end, and the same fate could befall a crew if their vessel lay becalmed for many weeks. Some pirates were reduced to eating their leather satchels or even to cannibalism to survive.

SCURVY KNAVES

Because of their cramped, damp and often filthy living conditions sailing ships were notorious for disease. Pirate ships were the worst of all, often overcrowded with the extra crew needed to man any captured vessel and usually lacking any proper medical equipment.

In the case of earlier pirate ships, little or no provision was made for lavatories (called the 'heads' and usually situated in the bow of later ships). Countermeasures such as scrubbing the decks with vinegar and salt water did little to help. The situation was made even worse in most ships by the large numbers of rats, which spread disease as well as eating precious food supplies and damaging the wooden hull.

Pirates fell easy prey to such diseases as dysentery, malaria, scurvy, smallpox, tuberculosis and typhus. It has been estimated that half of all the seamen who sailed during the Golden Age of piracy in the early eighteenth century died of disease. Half of an entire pirate crew could be lost to disease on a long voyage.

CASUALTIES

When fighting broke out it was often intense and bloody, with horrific injuries from cannonballs, musketfire, cutlasses and knives. The first man to board another vessel was almost certain to be killed. If he survived he got the first choice of any weapons captured on top of his share of booty.

Few pirate ships carried a capable surgeon, in which case amputations would be carried out by the ship's carpenter, who had a set of suitable saws. Inevitably, any serious wounds usually led to death through infection. Pirates who survived amputation might be fitted with a metal hook (in the case of a missing hand) or a wooden peg (to replace a severed leg). One such amputee was the French privateer François le Clerc, nicknamed *Pied de Bois* ('wooden foot').

Superstitions

Because of the dangers that they faced, most pirates were very superstitious. They sought to protect themselves magically by various means, such as tattoos and earrings. Many also believed that they could raise a helpful breeze through the use of magic.

THE PIRATE CODE

Pirates throughout history placed themselves beyond the reach of the laws governing legitimate seafarers, but they still obeyed their own self-imposed codes. Without these it would have been impossible for them to function effectively at sea.

DEMOCRACY ON BOARD

During the Golden Age of piracy, crews ran their lives according to a remarkably modern democratic set of rules, in which all crew members had their say. Ownership of vessels was sometimes shared, and any booty taken was split up equally or according to agreed proportions. In many cases the rules also laid down how captains were to be selected, as well as how they could be removed from office. The first such codes were drawn up by the early buccaneers. The rules (or 'articles') were often written down and

Incompetence

Incompetent pirate captains did not last long: one crew got through a total of 13 captains in just two months.

were formally agreed before the start of a voyage, or they were drawn up immediately after mutineers had taken over a ship.

In the days of the buccaneers, such documents might be accepted as legal evidence in a court of law in the case of any dispute. Anyone who broke the code was punished by his shipmates.

ELECTION OF OFFICERS

Pirate captains dictated what happened in battle but on many ships other decisions were taken by a majority vote. Captains and quartermasters were elected by a simple show of hands and could be replaced in the same way (as happened when Bartholomew Sharp was elected captain in 1680).

Captains Morgan, Kidd and Blackbeard were all elected captain by their crews, and both Morgan and Kidd suffered the indignity of also being voted out of office by dissatisfied crews.

BARTHOLOMEW ROBERTS' PIRATE CODE

One pirate code that has been preserved to this day is that agreed by the crew of 'Black Bart' (Bartholomew Roberts). It is recorded overleaf.

1 Every man shall have an equal vote in affairs of the moment. He shall have an equal title to the fresh provisions or strong liquors at any time seized, and shall use them at pleasure unless a scarcity may make it necessary for the common good that a retrenchment may be voted.

2 Every man shall be called fairly to turn by the list on board of prizes, because over and above their proper share, they are allowed a shift of clothes. But if they defraud the company to the value of even one dollar in plate, jewels or money, they shall be marooned. If any man rob another he shall have his nose and ears slit, and be put ashore where he shall be sure to encounter hardships.

3 None shall game for money either with dice or cards.

4 The lights and candles shall be put out at eight at night, and if any of the crew desire to drink after that hour they shall sit upon the deck without lights.

5 Each man shall keep his piece, cutlass and pistols at all times clean and ready for action.

6 No boy or woman to be allowed amongst them. If any man shall be found seducing any of the latter sex and carrying her to sea in disguise he shall suffer death.

7 He that shall desert the ship or his quarters in time of battle shall be punished by death or marooning.

8 None shall strike another on board the ship, but every man's quarrel shall be ended on shore by sword or pistol in this manner. At the word of command from the quartermaster, each man being previously placed back to back, shall turn and fire immediately. If any man do not, the quartermaster shall knock the piece out of his hand. If both miss their aim they shall take to their cutlasses, and he that draweth first blood shall be declared victor.

9 No man shall talk of breaking up their way of living till each has a share of £1,000. Every man who shall become a cripple or lose a limb in the service shall have eight hundred pieces of eight from the common stock and for lesser hurts proportionately.

10 The captain and the quartermaster shall each receive two shares of a prize, the master gunner and bosun, one and one half shares, all other officers one and one quarter, and private gentlemen of fortune one share each.

11 The musicians shall have rest by right on the Sabbath Day only. On all other days by favour only.

DISCIPLINE

Discipline upon pirate ships was more relaxed than on naval vessels, but if a pirate stepped out of line the punishment he faced could be severe. Many pirate captains were noted for their vicious tempers and stories were told of crew members who were killed on the spot after daring to challenge their leaders. Captain Kidd once killed his gunner, William Moore, in an argument by hurling a bucket at his head.

PIRATE TRIALS

Sometimes pirates held their own trials to determine the guilt of accused crew members. Evidence was heard on both sides before a judgement was reached and the punishment decided. According to Daniel Defoe, pirates even held mock trials for their own amusement, enacting the possible fates that lay in store for them if they were caught.

PUNISHMENTS

On naval ships the usual punishment for relatively minor crimes (such as carrying a candle into the hold without a lantern or striking another crew member) was flogging, but pirates resented the frequent use of

The infamous pirate Captain Kidd kills his gunner after his authority is questioned.

the cat-o'-nine tails, as this was exactly what many of them had sought to escape from when deserting previous ships. Nonetheless, floggings did happen, sometimes with each member of the crew delivering one stroke of the lash.

Keelhauling was among the most severe punishments. It involved the guilty man being hauled under the hull by ropes and pulled up (half-drowned and with much of his skin scraped off on the barnacles that grew on the keel) on the other side of the ship. The process might be repeated three times.

If a crew member murdered one of his fellows he might be tied to the corpse and thrown overboard with it to drown (borrowed from the Royal Navy).

MAROONING

Desert islands have long been central to pirate folklore. If a pirate was guilty of a serious crime against his fellow crew members, marooning on a remote island was one of the punishments that he might suffer.

Marooning was a cruel punishment, the pirate being abandoned on a desolate, often waterless, island or sandbank and left to starve to death. Typically he was left with just a bottle of water and a gun with some powder and shot. As there was usually little chance of rescue by another ship, and it was unlikely that he could survive by finding fresh water and food, he might take the option of using the pistol on himself and thus escape a slow, lingering death. Marooning was usually reserved for crew members guilty of theft from their fellows or of desertion from their ship, although Alexander Selkirk (the model for the fictional Robinson Crusoe) actually requested it after quarrelling with other crew members. Others who might be similarly abandoned included captives of pirates and other seamen who had refused to join their ranks. Shipwrecked pirate crews might also find themselves marooned on a remote isle.

The pirate Edward England and two companions were marooned by shipmates but eventually they escaped from the island of Mauritius by building themselves a boat and sailing back to civilization.

Being marooned and left to die on a desert island or sandbank was a harsh punishment.

LIFE ASHORE

Pirates did not spend all their time at sea. Periodically it was necessary for them to put ashore to replenish supplies, or to find a safe anchorage where they could carry out essential maintenance.

CAREENING

The wooden hulls of pirate vessels were easily damaged by marine worms and quickly became encrusted with barnacles and weeds, which could

Careening was the means by which the undersides of the ships were cleaned.

lead to a reduction in speed of up to three knots. The only solution was to 'careen' the ship, which involved beaching the vessel in a remote bay or estuary and then tipping it onto first one side then the other to scrape the hull clean using a chisel-like adz. A small vessel took a week to clean, but a larger ship took longer. Favourite locations for such maintenance during the Golden Age of piracy included swampy river estuaries on the west coast of Africa, particularly where the water was too shallow for large vessels to follow.

RELAXATION

When not working on essential maintenance the crews would make the most of the opportunity for any pleasure to be had. Among the few Spartan pleasures open to them in even the most remote places was a pipe of tobacco. Smoking was not allowed on board ship because of the danger of fire (crew members chewed the tobacco instead).

PLEASURES OF THE PORT

Most ports were closed to pirates, but there were still places where they could land without the danger of being arrested by the authorities and could enjoy the opportunity to spend their ill-gotten gains.

Pirates were often welcome visitors on shore because of the money they brought with them. The presence of even the most notorious pirates might be tolerated if they had money to spend: even Blackbeard himself was welcomed by the governors of North Carolina, on the understanding that he continued to leave English shipping alone.

Gambling and alcohol

There was no shortage of people ready to relieve the pirates of their riches, among them gamblers, tavern-keepers and women. Many a pirate embarked on a spree the moment he landed, indulging freely in alcohol, gambling and the company of women (who were not allowed on board ship). Some were drunk almost as soon as they set foot on shore, and stayed drunk until they returned to sea. Not surprisingly, it was often not long before whole crews needed to go back to sea again to get more money.

A HERO'S WELCOME

Some pirates enjoyed the status of celebrities and were hailed as heroes on putting into port. The English sea rovers were rapturously received on their return home, as were the French corsairs. American-born pirate Thomas Tew had such success at sea that he was fêted by fashionable New York society.

PIRATE LAIRS

Over the centuries some parts of the world acquired reputations as pirate haunts. In classical times, several Mediterranean ports were infested with pirates, while later the Barbary coast of north Africa inherited this reputation. Subsequent eras saw the establishment of other notorious pirate lairs around the world.

TORTUGA

Around 1620 the island of Tortuga, off northwest Hispaniola (now Haiti and the Dominican Republic) became a refuge for early French buccaneers from the Spanish. They used the island as a base from which to attack Spanish shipping with the encouragement of French governors, as they kept the island out of Spanish hands. The Spanish destroyed the colony in 1654, but withdrew under attack from the English.

Subsequent English governors of Tortuga attracted English and French buccaneers. Over the next 20 years they prospered (though Tortuga passed to the French again in 1659). Northern Hispaniola became the French colony of Saint Dominique and eventually the buccaneers moved to Petit Gloave in southwest Saint Dominique to sell their plunder in the markets there.

PORT ROYAL

Port Royal, Jamaica, became a notorious pirate lair after Jamaica was captured by the English in 1655. Buccaneers were encouraged to come here from Tortuga to help protect the port, from which they raided the Spanish Main. Dozens of buccaneer ships sailed from Port Royal, which became a thriving, lawless town with a population of 6,000, including many pirates, cutthroats and whores.

After the buccaneers turned to piracy the English authorities in Jamaica decided to drive them out. The 1681 anti-piracy law persuaded many pirates to relocate to the Bahamas and elsewhere, but Port Royal remained a wild, lawless place. Its destruction by an earthquake and tidal wave in 1692 was seen by many as divine punishment for the immorality of the town, which some called 'the wickedest city on Earth'.

NEW PROVIDENCE

The island of New Providence in the Bahamas was the most notorious pirate lair during the Golden Age of piracy in the early eighteenth century. It was used by many infamous pirates, among them Samuel Bellamy, Benjamin Hornigold, Henry Jennings, Jack Rackham, Edward Teach (Blackbeard) and Charles Vane.

Just 155 square km (60 square miles) in size, New Providence had no permanent population until the pirates came in the late seventeenth century. After the buccaneers left Port Royal in the 1680s, many moved to New Providence, which was conveniently close to trade routes. The Spanish attacked the island in 1684, but more buccaneers arrived in 1698 after leaving Tortuga. Over 500 pirates and 20 ships were based there by 1717.

Port Nassau, on the northern coast, possessed a natural harbour that was too shallow for warships and easy to defend. The market there flourished, selling liquor, slaves and other plunder. The town of shacks and tents grew rapidly with the arrival of ship repairers, whores and others supplying the pirates' needs. In 1718, however, the British government sent former privateer Captain Woodes Rogers to clean the place up. After he hanged some of the pirates, many others retired or left the island.

OCRACOKE ISLAND

Many pirates found secluded spots where they could rest up between voyages, safe from attack. Blackbeard established such a base at Ocracoke Island off North Carolina, where his anchorage was protected by a maze of sandbars. Here, in September 1718, he held a wild week-long pirate party for his own crew and that of fellow-pirate Charles Vane. The revels included much drinking of rum, music and dancing with women brought in from local towns. It was also at Ocracoke, however, that Blackbeard was surprised a few months later by naval vessels and killed in his last battle.

MADAGASCAR

The island of Madagascar, 400 km (250 miles) off the coast of south-west Africa, became a haven for pirates plundering the Indian Ocean and Red Sea in the early eighteenth century. A wild, exotic place with secluded anchorages and no settled European population, Madagascar played host to such pirate captains as Christopher Condent, Edward England, Henry Every, William Kidd and Thomas Tew.

Pirate lore depicts the pirates of Madagascar revelling in a tropical paradise, mixing with local women and living like princes. They did, however, build a fort on St Mary's Island off the east coast, just in case they were ever attacked. By 1700, St Mary's was home to some 1,500 pirates from 17 ships. In reality, life on the island was somewhat harder; the infamous Edward England, for instance, supposedly died a beggar on Madagascar, and the last pirates on the island were reported living in squalor.

The end of piracy

The introduction of convoys and stronger naval patrols curbed piracy in the region and many of the pirates of Madagascar moved on or became farmers. By 1711 only a handful were still active.

BARATARIA BAY, GALVESTON AND CUBA

The island of Grand Terre in Barataria Bay, south of New Orleans, became a notorious pirate lair in the early nineteenth century and was used by Jean Lafitte and others. They sold any captured slaves and their other plunder in New Orleans. If they were attacked, they retreated into the bayous and alligator-infested lagoons nearby.

The last pirates of the Caribbean

The pirates were eventually driven out by the US navy but relocated to Galveston, Texas, further along the coast. Galveston became a thriving pirate town with a big market for pirate plunder. After attacks on US shipping the town was destroyed in 1820.

The pirates then moved to northern Cuba, where the Spanish tolerated them until diplomatic pressure from the USA persuaded them to act. These last pirates of the Caribbean became robbers on the land, raiding local plantations.

TREASURE

The popular image of pirate treasure is that of a heavy iron-banded wooden chest stuffed with gold and silver plates and coins, gems, jewellery and costly silks. In reality, such riches were the exception and pirate treasure could be much more varied.

As well as gold, silver and gems, pirates would carry off anything they could sell or use themselves. In the ancient world, treasure might range from gold and silver to cargoes of grain or amphorae (large storage jars) of wine or olive oil. Pirates of later eras seized cargoes of cloth, rum, sugar, tobacco, furs, baled cotton, ivory, spices, ore, wood or salted meat and helped

Famous hauls

The greatest treasure hauls ever taken included those seized by Thomas Tew (from an Arab vessel in the Indian Ocean in 1694) and by Henry Every (from two Arab treasure ships in the Indian Ocean in 1695). Tew's haul was worth some £100,000, while Every's men shared over £600,000 (equivalent to £150 million today).

Captain Kidd is one of the few pirates who is known to have buried his treasure to keep it safe.

themselves to such practical items as carpenters' tools, ropes, sails, weapons, clothing and medical supplies. Food and liquor was seized without hesitation.

Sometimes pirates would take permanent possession of the captured vessel itself, casting the crew adrift or abandoning them on shore (having relieved them of their necklaces, rings, snuffboxes and any other valuables). Alternatively, the captives were forced to man the oars or were held for ransom. If slaves were taken, they were sold in the markets of New Orleans or elsewhere.

PIECES-OF-EIGHT

On the Spanish Main the greatest hauls came from the heavily-laden treasure galleons taking the gold of the New World to Spain. Pirates dreamed of seizing a fortune in gold doubloons or silver pieces-of-eight, which became the standard currency of Caribbean pirates. A gold doubloon was equivalent to what an ordinary sailor earned in two months.

BURIED TREASURE

The tradition that pirates buried their treasure to keep it safe until they were ready to retrieve it is largely a myth created by popular novelists such

This gold ring and golden Spanish coin are typical of the treasure seized by New World pirates.

as Robert Louis Stevenson. In reality, pirates generally kept their treasure locked safely away on board ship until they returned home, when it would be shared among the crew according to agreed proportions.

Very rarely, however, pirates needing to conceal their treasure temporarily might choose to bury it in some isolated spot. Sir Francis Drake, for instance, is said to have buried gold and silver snatched from a mule train at Nombre de Dios on the Isthmus of Panama and to have left it there under guard until his ships could pick them up. Captain Kidd is one of the few other well-known pirates believed by historians actually to have buried treasure to keep it safe. Kidd is thought to have buried his treasure on Gardiner's Island off Long Island, New York, but it was probably all recovered by the authorities not long afterwards. Much later, Jean Lafitte, whose ultimate fate is shrouded in mystery, may also have buried the bulk of his ill-gotten gains at some unknown location.

Dead men tell no tales

According to pirate lore it was common practice for evil-minded pirates to murder members of their crew who helped to bury their treasure, their skeletons remaining to warn others of their likely fate if they were tempted to steal the treasure for themselves. This prevented anyone else finding out where the treasure was located.

Treasure maps

Fictional pirates often leave hand-drawn maps that indicate the whereabouts of buried treasure (where 'X' marks the spot). Real pirates valued good maps, but do not seem to have ever drawn maps giving the location of their treasure, however cherished the notion may remain in the popular imagination.

FABLED TREASURE HOARDS

Despite the lack of evidence for the burying of treasure by pirates, legends have grown up about fabulous hoards of undiscovered pirate gold.

Certain locations in the Caribbean and elsewhere have long been identified as the likely location of pirate treasure, even though relatively little has been found by treasure hunters.

GARDINER'S ISLAND

The treasure that Captain Kidd buried on Gardiner's Island, off Long Island, New York, was probably dug up by the authorities after his arrest, but rumours persist that the bulk of it is still there. Legends have grown up around this valuable hoard including the superstition that the gold may only be retrieved by three people digging, in silence, at midnight under a full moon.

COCOS ISLAND

At least three separate hoards of pirate treasure are supposedly buried on Cocos Island, off the coast of Costa Rica. Remote and inaccessible, the island

Blackbeard's treasure

Edward Teach (Blackbeard) said that only he and the Devil knew the whereabouts of his treasure. Legend suggests that it was hidden on Smuttynose, one of the Isles of Shoals in New Hampshire, where he spent his last months. Four bars of silver were found there in 1820.

appeared on few maps of the pirate era. Those who are said to have left treasure there include the pirate Benito Bonito.

THE MONEY PIT

Another tantalizing treasure site is Oak Island off Nova Scotia, the location of the so-called 'Money Pit'. Since 1795, treasure hunters have tried repeatedly to locate treasure worth millions supposedly hidden at the bottom of an elaborate shaft.

Clues include the finding of a ship's block and tackle and several links of gold or copper chain. Numerous attempts to recover the treasure from the shaft, which floods regularly, have failed and the digging of new shafts has served only to confuse the exact location of the original pit.

PIRATE BATTLES

Pirates tried to avoid pitched battles and generally sought out easy prey who would not fight back. If a potential victim chose to resist, this not only endangered the pirates and their ship but also risked damage to the target ship and its cargo, from which the pirates hoped to profit. If their prey fought back most pirates would break off their attack and go in search of a more easily intimidated target.

If pursued, pirates usually relied upon superior speed to escape enemies. The shallow draught of many pirate vessels also meant they could go where many larger warships could not follow. On many occasions, however, the pirates had no option but to engage in a fight, and these could be both bloody and intense.

HAND-TO-HAND

Pirate battles at sea were usually fought at close quarters. If, after firing warning shots from a cannon, their prey did not surrender, the pirates might fire a single broadside of all their guns before drawing close to the other ship and preparing to board. Grappling irons would be thrown to draw the two ships together so that the pirates could board the victim. The two crews would then engage in bloody hand-to-hand combat with pistols, short swords, axes and daggers. The defending crew might blockade themselves from boarders in the strongest part of ship and would usually fight until all hope was lost, as any crew that resisted a pirate attack knew they could expect little quarter.

LAND BATTLES

Pirates also fought on land and some were capable of launching savage, often well-organised attacks against substantial defences. The buccaneers and sea rovers of the Spanish Main notched up many notable victories on land in the Americas, usually at the expense of the Spanish, while later pirates similarly raided coastal settlements and forts on both sides of the Atlantic and in the South China Sea.

The Battle of Panama

The Welsh buccaneer Sir Henry Morgan became notorious for his daring attacks on Spanish colonies in the Caribbean, which culminated in the Battle of Panama in January 1670.

Under the terms of his privateering commission, Morgan was not allowed to use ships to attack Spanish cities, so he developed a policy of landing his forces and launching attacks from inland. This also meant that any plunder he took did not have to be shared with the government, which took a large share of any booty taken at sea.

Morgan had previously captured the Spanish town of Puerto Príncipe in Cuba by attacking from inland (though with heavy losses) and had also seized the

Sir Henry Morgan leading his forces in the Battle of Panama, 1670, which was fought inland.

Spanish treasure city of Porto Bello in Panama using similar tactics. This latter triumph was particularly impressive, as the port was defended by three large forts and a substantial garrison.

JUNGLE TREK

Morgan gathered a formidable force of some 2,000 buccaneers in 36 ships for the assault on the gold city of Panama. He captured Fort San Lorenzo and then led his men on a gruelling 16-day trek through the jungle to Panama. The Spanish were waiting and attacked the buccaneers with 600 cavalry. Morgan's men managed to hold them off with musket fire and then beat back an attack by Spanish infantry. The Spanish went into headlong retreat and the city, with its immense fortune in gold, belonged to Morgan. The capture of Panama was the last great buccaneering venture and it ranks as one of the biggest battles fought anywhere in the Americas prior to the eighteenth century.

BLACKBEARD'S LAST BATTLE

The last battle fought by the infamous Edward Teach (Blackbeard) was perhaps the fiercest of all his engagements. Having preyed on colonial American shipping for 18 months, and thinking he was safe from retaliation from the corrupt governor of North

Carolina, Blackbeard retreated to his lair on Ocracoke Island off North Carolina in summer 1718.

Unknown to the fearsome pirate, however, the governor of Virginia now saw his opportunity and ordered Lieutenant Maynard of the Royal Navy to lead an attack on the pirate base with the lightly armed sloops *Ranger* and *Jane*.

Dawn attack

Maynard and his 60 men attacked at dawn on Friday 22 November 1718, catching Blackbeard by surprise, with just 25 men on board his sloop the *Adventure*. Blackbeard hastily cut his anchor cable and managed to slip across a patch of shallow water. When the two sloops then ran aground, he turned and gave each vessel in turn a devastating broadside. Maynard's *Jane* lost 20 men killed or wounded, while the *Ranger* lost 10 men. Unfortunately for Blackbeard, the *Adventure* then ran aground also. The *Jane* was first to free itself on the rising tide and came alongside the *Adventure*. Blackbeard threw grenades onto the deck of the *Jane* but caused few casualties, as Maynard was keeping most of his men hidden below decks. Deceived into thinking that they had the advantage in numbers, Blackbeard and his men subsequently boarded the *Jane*, only to find themselves surrounded by men pouring out of the hold.

Decapitated

Blackbeard himself fought like a demon, battling with his pistol and cutlass against Maynard. He received 25 wounds before he was finally decapitated by a Scots sailor. Another 10 of the pirates were killed; the rest were taken prisoner. When Maynard returned to his own sloop *Pearl* he had Blackbeard's severed head hung from the bowsprit.

FIGHTING EDWARD ENGLAND

Detailed contemporary descriptions of battles with pirates are relatively rare, but one exception was the engagement that was fought on 27 August 1720 between two ships under the captaincy of the notorious pirate Edward England and the heavily armed British East India Company ship *Cassandra*, under Captain James McRae. The intensity and savagery of this battle were typical of many pirate engagements of the time.

Attempted boarding

Abandoned by the two other ships with which he had neared the Jamaican coast, McRae found himself outgunned by the two pirate vessels, one of which had 30 guns while the second, England's *Fancy*, had 34 guns. While one of the pirate ships engaged in an exchange of cannon-fire with the *Cassandra*, the other attempted to get close enough to board, but was kept at bay by repeated broadsides that shattered her oars. The loss of life from the cannon-fire of the pirates was heavy and after three hours most of the officers and crew on the *Cassandra's* quarterdeck were dead. A second attempt at boarding was foiled when the pirate vessel ran aground.

A narrow escape

Further broadsides killed more of the *Cassandra's* crew and McRae's men grew increasingly desperate, knowing the pirates would probably kill them all for daring to resist their attack. As the pirates closed to board again, McRae ordered the surviving crew members to take advantage of the smoke around the ship to slip unseen into the *Cassandra's* longboat or swim to safety ashore. When the pirates finally got aboard all they found were three wounded men, who were immediately cut to pieces. McRae, who had been wounded, and the other survivors eventually surrendered. Against the wishes of his crew, England allowed them to go on to Kingston, Jamaica.

THE END OF BARTHOLOMEW ROBERTS

Bartholomew Roberts was another infamous pirate whose career was fated to end in a bloody sea battle with a Royal Navy ship. In his case, it came off west

Lenient pirates

Not all pirates committed atrocities. Because of the leniency that England showed to his prisoners, he lost the trust of his crew, who subsequently stripped him of his rank as captain and marooned him on the island of Mauritius.

Africa on the morning of 10 February 1722, when his ship the *Royal Fortune* was finally tracked down by *HMS Swallow*, under Captain Challoner Ogle.

Man-of-war

The *Swallow* was no ordinary naval vessel but a powerful man-of-war which, with 50 cannons, could outgun most pirate ships, including the *Royal Fortune*. It was part of a Royal Navy force that had been sent to intercept Roberts, whose piratical activities, it had been decided, could be tolerated no longer. By early 1722, Roberts had become one of the most successful pirates to sail the seven seas, having seized more than 200 ships in the course of his short career.

Audacious

The Navy found the pirates at anchor off Cape Lopez (in modern Gabon), recovering after a night spent celebrating a run of successes the previous day.

A handsome, flamboyant character, Roberts ordered his crew to action, intending to sail past the *Swallow* in an audacious bid to escape to the open sea.

As the two ships drew alongside the *Royal Fortune* was swept by a broadside fired by the *Swallow* at point blank range. Roberts was killed instantly, shot in the neck by grapeshot. His crew refused to give up but carried on the unequal battle for another three hours until they finally gave up and were taken prisoner. Before they surrendered they threw their dead captain overboard so that his body would not fall into his enemies' hands. The surviving pirates were put on trial and 54 of them were executed at Cape Coast Castle in west Africa.

Bartholomew Roberts' career ended when the Royal Naval warship *Swallow* attacked his ship.

PIRATE CAPTIVES

Captives faced an uncertain future. Some pirates were merciful and, after stripping captives of their valuables, put them ashore or entertained them as guests until a ransom was paid. Captured seamen might be invited to become pirates themselves. Other pirates, however, tortured captives (hoping they would reveal where treasure was hidden) before killing them.

Many pirates deliberately built up reputations for ruthlessness, knowing this encouraged future victims to cooperate. Some inflicted pain for pleasure and showed inhuman indifference to their victims' suffering.

FIERCE REPUTATIONS

The buccaneers had a particularly fierce reputation. Francis Drake had a deep hatred of the Spanish, and this was shared by many other English seafarers, as well as by French buccaneers. It was even known for Spanish captives to be put on the rack to extract the whereabouts of hidden treasure.

Barbary corsairs and Muslim pirates were greatly feared by their Christian counterparts and both sides showed little mercy, killing their captives or making

Admiral Trowbridge, barbarously wounded, is put in irons and then spiked to the deck by pirates.

them galley slaves. Chinese pirates in the South China Sea were similarly dreaded for their ill-treatment of prisoners, who were often tortured and beheaded.

Cruelty

The most infamous pirates of the Golden Age of piracy were noted for their cruelty. Stories were told of prisoners being shot, cut to pieces, hanged, burned, mutilated, keelhauled, marooned or (though rare in real life) forced to walk the plank. Others let their captives starve to death or, in the case of one pirate captain who captured a cargo of slaves, freeze to death as they sailed round the Cape of Good Hope.

Other torments included flogging, beating, 'sweating' (making a captive run round the mast until exhausted) and 'woolding' (tying cord round a prisoner's head and setting it alight). Pirates paid little respect to the rank or station of captives: captains had the most to fear, as pirates might seek revenge for past punishments.

RANSOM AND EXTORTION

Many pirates sought to make money from captives. If a rich or important person fell into their hands, they would demand a ransom for their release. Perhaps the most notable was Julius Caesar, who was held for five weeks by pirates as a young man; after his release he tracked the pirates down and executed them all.

Demands with menaces

Pirates also sought to make money through extortion, threatening the inhabitants of coastal settlements that they would return and destroy their towns unless money was handed over. Often hostages would be taken to reinforce the threat. Sir Henry Morgan threatened the destruction of whole cities unless he was paid a substantial ransom. Later pirates to follow his example were the buccaneers who plundered Cartagena in 1689 and the infamous Blackbeard who, in 1718, ransomed several prominent citizens of Charleston, South Carolina, and blockaded the port.

PIRATE ATROCITIES

Countless tales have been told of the atrocities that were committed by pirates over the centuries, but a small number of names stand out even among this catalogue of cruelties.

Few pirates could match the brutality of the Dutch privateer Rock Braziliano, a psychotic pirate captain operating in the Caribbean in the 1650s. He treated his Spanish prisoners barbarously, typically cutting off their limbs or roasting them alive over a fire, like pigs.

The Pirates striking off the arm of Capt. Babcock. p. 42.

Pirates had fearsome reputations; these are striking off the arm of a captured sea captain.

The French buccaneer François L'Olonnais also singled out the Spanish for brutal treatment, on one occasion cutting the heart out of a captive and then forcing another to eat it, still beating. Montbars of Languedoc, meanwhile, would nail one end of a captive's intestines to a tree and would then force the victim to dance until his guts were pulled out of his body and he died.

Some pirates were renowned for their cruelty and captives were sometimes thrown overboard.

Charles Gibbs

In terms of numbers killed, few outdid American pirate Charles Gibbs, who may have been responsible for the deaths of over 400 people.

Captives of Bartholomew Roberts might have their ears cut off, or be tied to a yardarm and used for musket practice, while later pirate Pedro Gibert was among those who locked crews below decks and set fire to their vessel. Edward Low was especially notorious: when one captain was intercepted off St Kitts in 1724 and dropped his gold into the ocean rather than let Low have it, the pirate cut off the man's lips and cooked them in front of him before murdering all the crew.

TREATMENT OF WOMEN

Atrocities against women captives were common. Although allegedly married 14 times, Blackbeard strangled most of the women he captured and threw their bodies overboard. Other pirates, seeing women as an encumbrance, simply threw them over the side to drown. In contrast, some pirate captains expressly forbade their men from molesting women; Captain John Phillips, for instance, threatened death to anyone who 'meddled' with women prisoners.

PIRATES AND SLAVERY

The huge profits that were to be made from the trade in slavery not surprisingly attracted succeeding generations of pirates. In the classical era, pirates often made slaves of their captives and put them to work in their galleys. Later pirates made money by selling captured slaves themselves.

SLAVERY IN THE NEW WORLD

In the sixteenth century, Spanish colonists in the New World began to import African slaves to work in their silver mines. In response to their demands, ships crossed the Atlantic laden with hundreds of African captives who were shackled in the hold. Because of the dreadful conditions in which these slaves were

Slaves as pirates

Occasionally pirates took pity on the slaves they captured and allowed them to join their crews (usually as servants, however, because they lacked seafaring experience). Some eventually became full members of the crew, earning an equal share of booty.

crammed together disease spread easily and many died on the voyage. Enough survived, however, to guarantee the slavers enormous profits. The first of many English seamen to sail as a slave trader (or 'picaroon') was privateer John Hawkins.

THE TRIANGULAR TRADE

The Golden Age of piracy in the early eighteenth century coincided with the peak in the so-called 'triangular' slave trade. This involved transporting slaves captured in Africa to the Americas, then taking goods bought with the proceeds of their sale to England to buy cheap goods that could in turn be exchanged in Africa for more slaves.

The slave ships made attractive prizes for pirates. Bartholomew Roberts, for instance, once captured 11 slave ships at Whydah (now Ouidah in Benin). The captured slaves could be sold by the pirates themselves to plantation owners in the markets of the Caribbean. Another ruse was to raid a plantation and carry off the slaves to resell elsewhere.

The slave trade eventually petered out after 1815, when the Royal Navy began to intercept slave ships crossing the Atlantic. This coincided with the gradual demise of piracy in the area.

PIRATES AND SMUGGLING

Smuggling involves the secret importing of goods that are either prohibited by law or upon which duty (tax) should normally be paid. Many pirates included smuggling among their various other piratical activities because of the profits to be made. Jean Lafitte, for instance, operated as a pirate, slave trader and smuggler throughout his career, supplying the markets of New Orleans and elsewhere with all manner of plunder on which no duty was paid.

IMPORT DUTIES

Smuggling dates back as far as the first import duties and increased steadily as duties were raised. In the eighteenth century, duty had to be paid on such everyday commodities as brandy, rum, wine, coffee, tea, tobacco, chocolate, salt and spices coming into England. To counter smuggling, 'customs men' were paid to patrol the likely landing places on the coast. Many of these officials, however, could be bribed to ignore anything they saw.

Success as a smuggler depended largely on good local knowledge and having the contacts on shore to receive and pay for the goods supplied. Good seamanship was essential, as ships had to unload in shallow water or secluded coves, usually at night. Goods were landed in small boats or left in shallow water weighted with stones for later collection.

SMUGGLING GANGS

Some areas became notorious for smuggling, among them Dorset and the whole of the south coast of England. Gangs of smugglers would gather to carry off any goods that were landed.

Smugglers earned reputations as heroes among the local populace, although many were as ruthless as any seagoing pirate. Smuggling was a dangerous occupation. If caught, a smuggler could be transported to the American colonies for seven years.

Moonrakers

The smugglers of Wiltshire were known as 'moonrakers'. This resulted from an incident when they were caught trying to rake up kegs of liquor hidden in a stream. When questioned, they feigned stupidity, pointing at the moon reflected in the water and claiming they were trying to rake up 'that large cheese'.

MYTHS AND LEGENDS

Some of the most familiar details of pirate lore owe more to the imagination of such novelists as Robert Louis Stevenson than they do to reality. In his pirate adventure *Treasure Island* (published in 1883), Stevenson popularized such stereotypical aspects of pirate life as buried treasure, treasure maps, parrots, peg legs, eyepatches and the dreaded black spot (a threat of death conveyed by means of a black spot symbol on a piece of paper).

Not all of these had any real historical authenticity. Although eyepatches, peg legs and parrots were doubtless associated with individual pirates, they were not unique to them. It is only since Stevenson that they have become familiar pirate symbols. Treasure maps were pure invention; with the possible exception of Captain Kidd, pirates rarely, if ever, buried treasure and therefore had no need of such maps. The black spot, also, was apparently Stevenson's own idea.

WALKING THE PLANK

The tradition of pirates blindfolding their victims and at cutlass point making them walk a plank over the ship's side to a watery death is perhaps the best-known

Here a pirate is being forced to walk the plank.

of the many pirate traditions that may in fact be no more than myth. It is often said that the ritual was entirely unknown to real pirates, but it seems that very rarely one or two historical pirate captains, notably Stede Bonnet, did in fact treat their victims in such a manner. As late as 1829 pirates blindfolded the crew of the Dutch brig *Vhan Fredericka* and forced them to walk the plank into the sea with shot fastened to their feet.

Walking the plank

The practice of walking the plank may date back to classical times, when pirates in the Mediterranean disposed of their Roman captives by politely inviting them to walk home over the sea.

PIRATE TALK

Pirates in the movies have a distinctive vocabulary. Indeed, certain words have become almost uniquely associated with pirates, although many were simply the vocabulary of all seamen of the period.

Pirate vocabulary

Aarr!/garr!/yarr!	(A general exclamation)
All hands hoay!	All hands on deck!
Avast!/avast ye!	(Dutch) Pay attention!
Go on the account	Embark on a piratical cruise
Landlubber	Person who lives on land (and is thus clumsy at sea)
Picaroon	(From Spanish) Rogue
Davy Jones' locker	The bottom of the sea (where dead seafarers go)
Matey	Mate/comrade
Shiver me timbers!	Blow me! (from the effect on a ship of gunfire or going aground).
Swab	Ignorant person (a swab being a mop made from rope used to clean decks)
Sweet trade	The career of piracy

PIRATE SONGS

Pirates at sea had few amusements and music was important to them. Most crews included musicians who entertained their fellows with fiddle music, hornpipes and sea shanties (seamen's folk songs). Their services were highly valued and they received one-and-a-quarter shares of any booty taken.

Like other seamen, pirates sang as they performed routine tasks about their ships, singing in rhythm as they pulled up anchors or hauled on sails. Most songs were the same as those of seafarers elsewhere, but one song has remained unique to pirates:

Fifteen men on a dead man's chest,
Yo-ho-ho, and a bottle of rum!
Drink and the devil had done for the rest,
Yo-ho-ho, and a bottle of rum!

This most famous of pirate songs was actually an invention of Scottish novelist Robert Louis Stevenson in his classic pirate novel *Treasure Island* (1883). Dead Man's Chest is a real place, an island in the Virgin Islands, which Stevenson read about in the travel book *At Last: A Christmas in the West Indies* written by Charles Kingsley.

PART FIVE

The end of the pirates

The fortunes of pirates around the world rose and fell as the various authorities sought to curb their activities. Numerous anti-piracy campaigns were conducted over the centuries, although it was not until the nineteenth century that piracy on a significant scale was consigned to history. However, even today piracy still exists on the high seas, especially in politically unstable regions of the world.

A warning to other pirates

The bodies of captured pirates, such as Captain Kidd, were sometimes hung in chains as a warning to their fellows of the fate that could await them if they were caught.

THE END OF AN ERA

The fortunes of the pirates waxed and waned over the centuries. Piracy in the classical world thrived for hundreds of years before a Roman campaign temporarily rid the Mediterranean of pirates.

In medieval times, the efforts of the Hanseatic League of northern European ports to curb piracy had little effect, and subsequently the galleons of Spain continued to fall prey to buccaneers and sea rovers despite their size and heavy armament.

THE GOLDEN AGE OF PIRACY

Some 5,000 years of piracy culminated in the so-called 'Golden Age' of piracy in the early eighteenth century, when Edward Teach (Blackbeard) and Captain Kidd were among the most feared pirate captains. Their moment of glory was short-lived, however. Piracy hindered trade and pushed insurance rates up, leading to a clamour for government action.

Woodes Rogers drove the pirates out of the Caribbean, while the Royal Navy hunted down pirate crews in their lairs or on the open sea. Blackbeard and Bartholomew Roberts were among those killed in battle with Royal Navy ships. Captain Kidd, Jack Rackham, Charles Vane and Stede Bonnet had their careers cut short by the public hangman.

NAVAL CAMPAIGNS

It was another 130 years, however, before piracy was finally eclipsed. After the Napoleonic Wars ended in 1815 the Barbary corsairs were crushed by combined US and European navies; in 1816 British and Dutch ships bombarded the port of Algiers in retaliation for local tolerance of pirates, forcing an apology from the Algerians and the end of piracy in the region.

Other nests of pirates were destroyed by similar campaigns conducted by the navies of Britain, the USA and other maritime powers in Malaysia, Indonesia and elsewhere into the 1840s.

Steam ships

The new steam-powered vessels of the world's navies outclassed the sailing ships and junks of the pirates, and by 1850 their breed was nearly extinct. In 1856, privateering also ended under the Declaration of Paris, which stopped the issue of letters of marque under which privateers sailed.

PIRATE HUNTERS

There were many notable pirate hunters throughout the ages, the most famous of whom are listed here.

POMPEY THE GREAT

Pompey the Great (106–48 BC) was the Roman leader who conducted an ambitious campaign against the powerful Cilician pirates in the first century BC and brought about their downfall. He was asked to rid Rome of the pirate menace that by 67 BC had become the biggest pirate confederation the world had yet seen, threatening the delivery of grain to Rome by closing Mediterranean trading routes. Pompey amassed a huge fleet of 500 galleys and in a brilliant three-month campaign set about rounding up and defeating the pirates, who until then had enjoyed virtually unchallenged control of the sea lanes.

'Ruler of the Sea'

Ironically, Pompey's son Sextus (67–36 BC) turned to piracy to attack his rival Octavian, pillaging the Italian coastline from his base in Sicily and calling himself 'Ruler of the Sea'.

Executions

As many as 10,000 pirates were killed or executed and their base in Cilicia (in southern Turkey) was seized by the Roman army. From then onwards, the pirates of the Mediterranean remained a minor nuisance to the Romans, though they never disappeared.

PEDRO DE MENÉNDEZ DE AVILES

Pedro de Menéndez de Aviles was Spain's most successful pirate hunter in the sixteenth century, when Spanish treasure ships and possessions in the West Indies were falling prey to attacks by chiefly Protestant French Huguenot pirates.

The problem had become so serious by 1562 that the Spanish king Philip II appointed the nobleman Pedro de Menéndez de Aviles captain-general of the annual treasure fleet with orders to protect it from the pirates. As an owner of merchant vessels sailing to and from the New World, de Menéndez was familiar with the pirate threat and had already made a study of Spanish defences in the Spanish Main.

Victory against the French

De Menéndez took steps to secure the safety of the annual treasure fleet, but on his return to Spain was arrested and tried on charges of smuggling. The king, however, stepped in to rescue de Menéndez, whom he had entrusted with the task of removing the newly-established French Huguenot pirate base at Fort Caroline in eastern Florida.

In 1565, de Menéndez led 30 ships across the Atlantic and founded a new Spanish colony at St Augustine in Florida. When the French pirate fleet sailed from Fort Caroline he attacked the lightly-defended base and ruthlessly executed all the prisoners he took. He then dealt out the same fate to the 200 survivors of the French pirate fleet who limped back to shore after many of the ships were wrecked in a storm.

These successes largely ended the Huguenot interference with the Spanish treasure convoys. The Spanish king rewarded de Menéndez by making him governor of Havana.

Overcoming tribulations

De Menéndez experienced many difficulties before achieving his final victory over the pirates. Not least of these was a storm that scattered his 30 ships as they crossed the Atlantic in 1565. He was left with five ships and 600 men to accomplish his task.

WOODES ROGERS

Born in Poole, Dorset, around 1679, the pirate hunter Woodes Rogers went to sea as a boy and later had a successful career as a privateer captain and slave trader. He sailed round the world twice between 1708 and 1711 and made his fortune with treasure seized from the Spanish colonial port of Guayaquil in Ecuador and from the capture of the so-called 'Manila galleon' carrying gold from the Philippines to Spain.

Governor Woodes Rogers is shown here with members of his family.

Driving out the pirates

In 1717 Rogers was asked to curb piracy on New Providence in the Bahamas, which had become a notorious pirate lair. He sailed to Nassau on New Providence with a Royal Navy escort. By the time he arrived, in July 1718, many of the pirates had already left. The only serious resistance came from Charles Vane, who set a fireship ablaze and fired a cannon in mock salute as he sailed away.

The 600 pirates who remained accepted pardons and promised to give up piracy. Rogers installed a civil council and raised three militia companies to defend the island. Former pirates Benjamin Hornigold and Thomas Burgess were appointed official pirate hunters. Rogers also improved the harbour defences and rebuilt the harbour fort.

Mass hanging

After the hanging in December 1718 of eight men who had returned to piracy New Providence was never again threatened by pirates. A Spanish attack in 1720 was beaten off, but a year later, lacking money and government support, Rogers returned to England, was dismissed from his post, declared bankrupt and thrown into debtors' prison. His fortunes were later restored and in 1728 he was reappointed Governor of the Bahamas with proper powers and a salary.

COMMODORE DAVID PORTER

Commodore David Porter was a US Navy officer and became the most effective of the pirate hunters who were responsible for driving out the last generation of pirates to prey on shipping in Caribbean and American waters.

With some 2000 pirates still operating in the area in the early 1820s, despite the best efforts of US frigates and sloops under the leadership of such respected captains as Lieutenant Commander Lawrence Kearney, it was realized that more success might be had with smaller vessels which were capable of following the pirate ships into shallow water.

The West India Squadron

A special West India Squadron was therefore formed under the command of the brilliant but also fiery-tempered 42-year-old Commodore David Porter. Porter had first gone to sea as a midshipman in 1798 and over the intervening years had achieved the status of naval hero.

The 'Mosquito Fleet'

Porter assembled a fleet of eight shallow-draughted schooners, as well as an early paddle-steamer (the first steam-powered ship ever to engage in naval warfare) and five flat-bottomed oar-powered barges. To lure the pirates he also fitted out a decoy ship that looked like a merchant vessel, but which actually mounted six hidden cannons. The so-called 'Mosquito Fleet' was manned by 1,150 sailors and marines, and was based in Key West.

A big success

The squadron ruthlessly hunted the pirates down, destroying vessels and taking many prisoners. Early in 1823 Porter had his biggest success when he joined forces with six US warships and defeated the feared Cuban pirate Diabolito ('Little Devil') in his two-masted schooner *Catalina* off northern Cuba, resulting in the death of 70 pirates.

The end of piracy

Over the next two years Porter captured hundreds more pirates and effectively ended piracy in the West Indies. The West India Squadron itself, its job done, was disbanded shortly afterwards.

Broken careers

Among the many pirates whose careers ended as a result of Porter's efforts were the infamous Charles Gibbs and Jean Lafitte.

TRIALS AND EXECUTIONS

THE CRIME OF PIRACY

Anyone who committed robbery below the high-tide mark could be tried for piracy, an offence for which a convicted person risked the death penalty. Privateers, who claimed to act on behalf of their governments in time of war, often argued successfully that they were not guilty of piracy and could hope to be exchanged for other prisoners, though on many occasions their letters of marque proved inadequate legal protection. In any case, any person thrown into prison on charges of piracy might never get to trial as they might easily die of disease, the fate of many prisoners.

TRIALS

In Britain, pirate trials were conducted with the authority of the Admiralty. If the trial took place in one of Britain's colonies before 1700 it was presided over by the local colonial governor; after 1700 a special Vice-Admiralty court was set up, assisted by local dignitaries and any sea captains in the area.

Trials rarely lasted more than a day or two. Accused persons, however ill-educated, had to conduct their own defence. Many pirates offered no defence at all,

The hanging of a pirate at Execution Dock, London. Executions often attracted large crowds.

though some tried to excuse their crimes on the grounds that they were drunk at the time, or else had been forced to serve in pirate crews against their will. Such claims were often hard to prove. Pirate trials and the punishments that followed were well-publicized to deter others from following a similar path.

SENTENCING

Pirates through the ages knew that if they were tried and found guilty of piracy then they were very likely to be condemned to death. In the seventeenth century, it was usual for only captains and other ringleaders of pirate crews to be hanged, while the rest of the crew were given lesser punishments.

In the eighteenth century, however, entire pirate crews might receive the death sentence. Lesser punishments for piracy included long terms of imprisonment (often death sentences in themselves because of the filthy, overcrowded conditions in most prisons) or transportation to a distant prison colony.

PIRATE TRIALS

Some pirate trials were better-organised affairs than others. Often some attempt was made to give the accused men a fair hearing, but in practice accused pirates had little chance of clearing their names. The authorities preferred guilty verdicts as these could be used as a warning to other pirates still at large.

Trial of Captain Kidd

Among the most sensational of all pirate trials was that of Captain William Kidd and nine of his crew before six judges at the Old Bailey in London in May 1701. Confined in Newgate prison for the past year, Kidd was in poor health and had already made a bad impression when called to testify to the House of Commons about the possible involvement of leading Whig politicians in his piratical activities.

In court, denied the chance to consult with legal advisers until the last moment, he was unable to furnish a defence to the charge that he had murdered one of his crew and failed to discredit the evidence of two former crew members who had deserted him in Madagascar. With no witnesses of his own, he could only accuse the Crown's two witnesses of perjury and resign himself to the inevitable death sentence.

Trial of Black Bart's crew

Another notable pirate trial was that of the crew of Bartholomew 'Black Bart' Roberts, which took place before a Vice-Admiralty court in west Africa in 1722. Of the 268 men captured, 77 were black Africans who were sold into slavery instead of being tried. Nineteen men died in prison of wounds or disease before they could appear in court.

When the proceedings ended 52 of the pirates were sentenced to hang. Another 37 were imprisoned, while 20 were sentenced to seven years' hard labour in African gold mines and 17 were imprisoned in Marshalsea Prison. Those who claimed to have been forced into piracy against their will were acquitted.

PIRATE EXECUTIONS

During the medieval period convicted pirates might be beheaded, as were those in the South China Sea, but otherwise pirates usually faced death by hanging or, in Spain, by the garrotte (a particularly unpleasant method of strangling to death).

In Britain, convicted pirates were hanged below the high-tide mark in acknowledgement of the fact that crimes committed below that point fell under the Lord High Admiral's jurisdiction, not that of the civil courts. Many notable English pirates were hanged on the banks of the Thames at Execution Dock in Wapping, London.

Mass executions

Mass executions of pirates were sometimes staged to deter others. In 1573, for instance, German pirate Klein Henszlein and his entire crew were beheaded in Hamburg; by the end the executioner was said to be ankle-deep in blood.

Captain Kidd, tarred and hung in chains, to warn other pirates of what would befall them if caught.

Pirate hangings would attract large crowds though some pirates met their end more bravely than others. On approaching the gallows in Charleston harbour elegant pirate Stede Bonnet broke down completely and women in the crowd were reportedly much affected by the pathetic figure he made. Others, however, went cheerfully to their deaths or drank so much that they barely knew what was happening.

HANGING IN CHAINS

The body of a dead pirate would be left until three tides had washed over it. It would then be buried in an unmarked grave or otherwise tarred and hung in chains (an iron cage suspended from a gibbet) at a prominent location as a warning to other pirates. Captain Kidd was among those whose bodies were displayed like this, as were Charles Vane and Jack Rackham. The body would remain until it rotted away.

Pirates were said to dread this fate even more than being hanged. Alternatively, the body was handed over for dissection by anatomists.

EXECUTIONS OF PIRATES 1700–30

May 1701 24 Frenchmen from the pirate ship *La Paix* hanged at Execution Dock, London.
Captain Kidd and eight men from the *Adventure* Galley hanged at Execution Dock, London.

June 1704 Captain Quelch and six men from the *Charles* hanged on the Charles River, Boston, Massachusetts.

Summer 1705 Captain Green and 16 men from the *Worcester* hanged on Leith Sands, Edinburgh.

November 1716 Captain Dolzell hanged at Execution Dock, London.

June 1717 The pirates De Mont, De Cossey, Rossoe and Ernandos hanged at Charleston, North Carolina.

October 1717 Six of Samuel Bellamy's crew hanged on the Charles River, Boston, Massachusetts.

March 1718 13 men from Edward Teach's (Blackbeard's) sloop *Adventure* hanged at Gallows Road, Williamsburg, Virginia.

October 1718 Stede Bonnet and 29 of his crew hanged in Charleston harbour, Charleston, North Carolina.

December 1718 Eight pirates hanged in Nassau, New Providence, Bahamas, by Woodes Rogers.

February 1719 Captain Worley and one other hanged at Charleston harbour, Charleston, North Carolina.

November 1720 Captain Rackham and nine others hanged at Gallows Point, Spanish Town, Jamaica.

November 1721 Charles Vane and one other hanged at Gallows Point, Spanish Town, Jamaica.

March 1722 52 of Bartholomew Roberts's crew hanged at Cape Coast Castle, West Africa.

May 1722 Italian pirate Captain Luke and 40 men hanged at Gallows Point, Spanish Town, Jamaica.

October 1722 Five of the crew of Spanish pirate Captain Blanco hanged at Nassau, New Providence, Bahamas.

Summer 1723 Captain Finn and four pirates hanged in St John's harbour, Antigua.

July 1723 Captain Massey hanged at Execution Dock, London.
Captain Harris and 25 crew of the *Ranger* hanged in Newport harbour, Newport, Rhode Island.
Captain Philip Roche hanged at Execution Dock, London.

March 1724 11 pirates from Captain Lowther's crew hanged in St Kitts.

May 1724 Captain Archer and one other hanged on the Charles River, Boston, Massachusetts

May 1725 Captain John Gow and seven crew of the *Revenge* hanged at Execution Dock, London.

July 1726 Captain William Fly and two pirates hanged on the Charles River, Boston, Massachusetts.

July 1727 Captain John Prie hanged at Execution Dock, London.

PART SIX

The legacy of pirates

The Golden Age of piracy may be long gone, but the influence of the pirate upon popular culture has never been stronger. Images of pirates are now commonplace throughout advertising, fiction and the media. The modern image of the pirate remains highly romanticized, despite the efforts of historians to correct this largely misleading impression.

Long John Silver

The fictional pirate Long John Silver in Robert Louis Stevenson's famous book *Treasure Island* is the archetypal pirate of literature and has helped to reinforce many myths about pirates.

PIRATES IN LITERATURE

The long-cherished tradition of the swashbuckling pirate hero that is so firmly entrenched in the popular imagination had its roots in literature. The classic image of the pirate was first presented in Alexander Exquemelin's book *Buccaneers of America*, published in English in 1684. Frenchman Alexander Exquemelin (1645–1707) was a former buccaneer and had met Sir Henry Morgan among other significant figures. His book enjoyed huge success and established the pirate genre in entertainment.

A POPULAR GENRE

A play entitled *The Successful Pirate* (1713) brought to the stage the adventures of English pirate Henry Every and continued the romanticization of the subject. It was followed in 1724 by Captain Charles Johnson's *A General History of the Robberies and Murders of the Most Notorious Pirates*, which remains an important source of information about pirates of the early eighteenth century. The identity of Charles Johnson has been the subject of much speculation. Some have suggested that he was merely a pseudonym for the novelist Daniel Defoe, who wrote several books on pirates.

ROBINSON CRUSOE

Daniel Defoe's masterpiece *Robinson Crusoe* (1720) vividly described the adventures of a sailor marooned on a desert island. Crusoe was based on a real seaman, Alexander Selkirk (1676–1721), who sailed under privateer William Dampier before being put ashore (after quarrels on board ship) on Más a Tierra, one of the Juan Fernández islands in the South Pacific (transformed by Defoe into an island near Trinidad), in 1704. He remained there for five years until he was rescued by Woodes Rogers.

TREASURE ISLAND

The theme of pirates continued in popular fiction into the nineteenth century. Lord Byron's poem *The Corsair* (1814) and Sir Walter Scott's novel *The Pirate* (1821) were both highly successful, but even more influential was Robert Louis Stevenson's classic

Musical pirates

Pirates have also inspired several musical works over the years, among them Giuseppe Verdi's opera *Il Corsaro* (1848) and Gilbert and Sullivan's *The Pirates of Penzance* (1879).

adventure story *Treasure Island* (1883). Stevenson's tale recounted the adventures of young Jim Hawkins among pirates in search of buried treasure and introduced the archetypal one-legged pirate Long John Silver, complete with parrot. Stevenson's book introduced or reinforced such myths of piracy as buried treasure and treasure maps.

PETER PAN

J.M. Barrie's story *Peter Pan*, featuring the dastardly Captain Hook and his pirate crew, began life as a novel and was first produced as a stage play in London in 1904. Barrie's tale added new details to the pirate myth, including the stereotypical hook for a hand and hats decorated with a skull-and-crossbones symbol. Captain Hook himself was loosely based on real-life pirate Edward Teach (Blackbeard).

Modern literature

Piracy of the swashbuckling variety remained a popular choice of subject among romantic adventure novelists, such as Rafael Sabatini (1875–1950) well into the twentieth century. Many of their stories are familiar to modern audiences through various cinematic adaptations. Today pirates are stock characters in a wide range of entertainments, from films and novels to children's cartoons.

"HE WAS BRAVE AND NO MISTAKE"

Robert Louis Stevenson's one-legged Long John Silver, the archetypal pirate in classic literature.

Peter Pan battles Captain Hook in an illustration from the popular children's book *Peter Pan*.

FILM PIRATES

The first film to feature pirates was the silent movie *Treasure Island*, released in 1920 and based on Robert Louis Stevenson's novel. Ever since then pirates have made numerous appearances on the silver screen, particularly in the 1940s and 1950s, and it is chiefly through films that modern generations have become familiar with pirate history and lore.

Taking their cue from popular fiction, film-makers have taken numerous liberties with reality, changing historical events, contriving meetings between pirates who lived in different eras and adding further elaborations to pirate mythology. In particular, alongside the vicious cutthroats of authentic history, cinema has developed the notion of the dashing swashbuckling buccaneer or pirate hero, a cunning rascal with a irresistible romantic appeal.

FILM BIOGRAPHIES

Several attempts have been made to portray the life stories of famous pirates on the big screen, though usually without historical authenticity. Examples have included *Captain Kidd* (William Kidd), *Seven Seas to Calais* (Sir Francis Drake), *The Buccaneer* (Jean Lafitte),

Morgan the Pirate (Henry Morgan), *Anne of the Indies* (Anne Bonny and Blackbeard) and *Blackbeard the Pirate* (Blackbeard and Henry Morgan).

PIRATE ACTORS

Many notable movie actors have played pirates, like Douglas Fairbanks Senior, the original cinematic swashbuckling hero-pirate, Charles Laughton, Errol Flynn (who escaped pursuers by plunging his dagger into a sail and then sliding down to the deck as it sliced the canvas), Robert Newton (a memorable Long John Silver), Tyrone Power, Burt Lancaster, Dustin Hoffman (as J.M. Barrie's Captain Hook), and Johnny Depp (as Jack Sparrow) and Geoffrey Rush (as Captain Barbossa) in *Pirates of the Caribbean*.

PIRATE FILMS

There have been many classic pirate films, including the following: *Treasure Island* (1920, 1934, 1950, 1971, 1990, 1991); *Captain Blood* (1924, 1935); *The Buccaneer* (1938, 1958); *Swiss Family Robinson* (1940, 1960); *Blackbeard the Pirate* (1952); *A High Wind in Jamaica* (1965); *Hook* (1991); *Muppet Treasure Island* (1996); *Cutthroat Island* (1997); *Pirates of the Caribbean: The Curse of the Black Pearl* (2003); and, more recently, *Pirates of the Caribbean: Dead Man's Chest* (2006).

PIRATE WRECKS

Most pirate ships did not last long. If not destroyed in battle, they might be sunk in storms, run aground, burned, or lost through accidents or carelessness. In many cases, they were simply abandoned because they were no longer seaworthy. This was particularly true of the ships sailed in the Caribbean, where the wooden hulls suffered extensive damage from holes bored by teredo worms. Regular cleaning (careening) of the hull prolonged the ship's life, but sooner or later it would become too rotten to keep water out.

If a ship had to be abandoned, the pirates would usually strip the vessel of anything that could be sold or used on a replacement ship. What was left might then be burned to the waterline. This seems to suggest that wrecks found today have little to offer, but in reality much can be learned from what remains.

WRECK HUNTING

Tales of fortunes in treasure from sunken pirate ships might seem the stuff of fantasy, but objects of great value have been salvaged, such as gold and silver coins, jewellery and gold dust, as well as historically important cannons and personal belongings.

Locating wrecks

Wreck hunter Barry Clifford is one of the leading names in the location of pirate wrecks, having found (among others) Samuel Bellamy's *Whydah* and Captain Kidd's *Adventure Galley*. It took Clifford 15 years to find the *Whydah* and another 20 years to bring the finds to the surface. Items salvaged from the wrecks have contributed much to contemporary knowledge about the pirate ships of the early eighteenth century.

The Whydah

The wreck of the pirate galley *Whydah* was located off Cape Cod in 1984 and subsequently over 100,000 objects have been salvaged from the site, providing much information about pirate ships of the early eighteenth century. The *Whydah* was a British slave trader until captured by the English pirate Samuel Bellamy early in 1717. The 272-tonne (300-ton) *Whydah* was then fitted out with 28 guns and preyed on shipping off the Atlantic coast of the American colonies. During the night of 17 May 1717, however, the *Whydah* struck a sandbar in heavy fog and capsized, with the loss of all but two of her crew of 146 men. Bellamy was among those drowned. Finds from the wreck have included the ship's bell, weapons, coins, eating utensils and 100 pieces of African gold jewellery.

Queen Anne's Revenge

Perhaps the most sensational pirate wreck found to date is that of *Queen Anne's Revenge*, the flagship of the infamous pirate Edward Teach (who was better known as Blackbeard), which was located in 1996.

Formerly a French slave ship called *La Concorde*, the vessel was captured by Blackbeard in late 1717. Renamed and rearmed, the ship then terrorized the Atlantic coastline of the American colonies before finally running aground on a sandbar in Beaufort Creek, North Carolina.

It is thought that Blackbeard may have run the ship aground deliberately in order to reduce the size of his crew, many of whom he then marooned (although the position of a kedge anchor on the seabed suggests that attempts were made to pull the ship free). Finds from the wreck have included a bell, some gold dust, platters and cannons forged in England.

Adventure Galley

The wreck of Captain Kidd's *Adventure Galley* was located in a bay off north-eastern Madagascar by wreck hunter Barry Clifford in 1999. Kidd had set fire to the ship, which had a rotten hull, in 1698 and left it to sink in shallow water. Finds from the site have included English oak timbers and pewter objects.

MODERN PIRACY

Although piracy is usually thought of as belonging to the past, it has never completely gone away, flaring up from time to time chiefly in unstable regions of the world. Reports of pirates operating in significant numbers in the South China Sea, off east Africa and in Brazilian coastal waters have increased in number since the 1980s. These modern-day pirates attack tankers and other large ships as well as luxury yachts and merchant vessels, using fast, small boats and threatening crews with AK-47 automatic rifles, grenade launchers and other weapons.

Pirates still operate in the South China Seas, but they now use sleek, fast powerboats.

Luxury liner under fire

In November 2005, piracy hit the headlines again when pirates attacked the luxury liner *Seaborne Spirit* in the Seychelles, firing rocket-propelled grenades and automatic gunfire from a small boat. The liner outran the pirates and on this occasion only one crew member was hurt.

The pirates come in search of any loot they can plunder, although occasionally the ships themselves have been seized. Some captured crews have been maltreated or thrown overboard. The pirates' activities also include smuggling drugs and contraband goods. Many of them are fishermen, though some belong to criminal gangs, regional independence movements or even to local government forces.

ATTACKS

A Piracy Reporting Centre was set up in Kuala Lumpur, Malaysia, in 1992. Piratical activity reached a recent peak in 2004, when pirates were held responsible for the deaths of 30 people (up from 21 in 2003). In 2004, 325 pirate attacks were reported worldwide, of which 93 took place in the waters of Indonesia and 39 off Nigeria.

FIND OUT MORE

PIRATE BOOKS

A General History of the Robberies and Murders of the Most Notorious Pirates, Captain Charles Johnson, 1724 (and later eds.)
A History of Pirates, Nigel Cawthorne, 2005
A Nation of Pirates: English Piracy in its Heyday, Clive Senior, 1976
Between the Devil and the Deep Blue Sea, Marcus Rediker, 1987
Blackbeard the Pirate, Robert E. Lee, 1974
Bold in her Breeches, Jo Stanley, 1995
Captain Kidd and the War against the Pirates, Robert C. Ritchie, 1986
English Corsairs on the Barbary Coast, Christopher Lloyd, 1981
Piracy Today, Roger Villar, 1985
Pirates 1660–1730, Angus Konstam, 1998
Pirates: Adventurers of the High Seas, David F. Marley, 1995
Pirates! An A–Z encyclopaedia, Ian Rogozinski, 1995
Pirates and Privateers, David J. Starkey, 1997
Pirates Fact and Fiction, David Cordingly and John Falconer, 1992
Pirates of the Caribbean: Buccaneers, Freebooters and Filibusters, Cruz Apestegui, 2002

Pirates of the West Indies, Clinton V. Black, 1989
Pirates: Terror on the High Seas from the Caribbean to the South China Sea, 1996
The Black Ship: the quest to recover an English pirate ship and its lost treasure, Barry Clifford, 1999
The History of Piracy, Philip Gosse, 1990
The History of Pirates, Angus Konstam, 2005
The Pirate Prince: Discovering the Priceless Treasures of the Sunken Ship Whydah, Barry Clifford and Peter Turchi, 1993
The Pirate Ship 1660–1730, Angus Konstam, 2003
The Pirates, Douglas Botting (ed.), 1978
Treasure Wreck: the fortunes and fate of the pirate ship Whydah, Arthur T. Vanderbilt, 1986
Under the Black Flag, David Cordingly, 1995

PIRATE WEB SITES

The following are a brief selection of pirate websites:
http://www.ah.dcr.state.nc.us/qar/
(information about the wreck of *Queen Anne's Revenge*)
http://www.geocities.com/Athens/7012/index.htm
(biographies and other information)
http://www.geocities.com/Tokyo/Garden/5213/index.htm (general information)
http://www.nationalgeographic.com/pirates/
(games and information)
www.piratesinfo.com/ (general information)

GLOSSARY

Abaft To the rear of.
After Towards the stern of a vessel.
Barque A sailing ship with fore-and-aft rigged sails.
Becalmed Unable to move in windless conditions.
Block and tackle A system of ropes and pulleys used to lift heavy loads and keep ropes taut.
Booty Gold, silver and other treasure.
Bow The forward end of a vessel.
Bowsprit A sail-carrying spar projecting forward from a vessel's bow.
Brace A rope controlling horizontal movement of a square-sailed yard.
Brig A two-masted, square-rigged vessel with a fore-and-aft sail abaft the mainmast.
Brigantine A two-masted vessel with a square-rigged foremast and fore-and-aft rigged mainmast with a square-rigged topsail.
Broadside The simultaneous firing of all the cannons on one side of a vessel.
Buccaneer A pirate or privateer who plundered Spanish ships and coastal towns in the Caribbean.
Bulkhead A vertical partition within the hull of a vessel.
Caravel A medieval sailing ship with a high poop deck used by the Portuguese and Spanish in the fifteenth and sixteenth centuries.
Careening Scraping a beached vessel's hull clean of weeds and barnacles.
Carrack A medieval trading vessel resembling an early galleon.
Cat-o'-nine tails A whip with nine separate strands of rope.
Caulk To fill the seams between a vessel's planks with rope and hot pitch.
Chain shot A form of cannon ammunition

comprising two metal balls linked by chain and used to destroy masts and sails, etc.

Chart A map.

Colours Flags identifying a ship's nationality.

Corocoro A fast sail or oar-powered vessel with outriggers used by Indonesian pirates.

Corsair A pirate or privateer who operated in the Mediterranean.

Crow's nest A lookout platform high up on a mast.

Cutlass A short sword used by buccaneers and pirates in close combat.

Cutter A small single-masted vessel with a fore-and-aft rigged mainsail, foresail and jib.

Deadeyes A round wooden block to extend shrouds.

Displacement The amount of water displaced by a vessel's hull.

Doubloon Spanish gold coin equivalent to 16 pieces-of-eight.

Draught The depth of a loaded vessel in the water.

East Indiaman A large English or Dutch merchant vessel sailing between Europe and Asia.

Fathom A depth of six feet of water.

Filibuster A French buccaneer and, later, a smuggler or blockade runner.

Flagship a ship carrying an admiral or the senior commander of a fleet of vessels.

Flintlock pistol A pistol fired by means of a spark struck by a flint when the trigger is pulled.

Fluyt A shallow-draughted Dutch merchant vessel, usually with three masts and square sails.

Fore-and-aft rigged A vessel with the sails set in line with the hull.

Forecastle A raised deck towards the front of a vessel, otherwise called the 'fo'c'sle'.

Foremast A mast at the front of a vessel.

Freebooter A pirate or other person living by plunder.

Gaff-rigged A vessel with a fore-and-aft rigged sail on a fore-and-aft yard mounted on a vertical mast.
Galiot A form of small galley.
Galleon In the sixteenth and seventeenth centuries, a large three-masted ship with square-rigged sails.
Galley A large oar-powered ship.
Gallows A wooden structure for hanging condemned criminals.
Gibbet A wooden structure used to display the bodies of hanged criminals.
Grappling iron Metal hook attached to a rope, tossed onto another vessel to haul it close enough for boarding.
Guineaman A slave ship.
Gunwale The upper planking of a vessel's side.
Halyard A rope used to hoist a colour or sail.
Hardtack A hard and dry ship's biscuit.
Heave-to To cease moving.
Helm A vessel's tiller or wheel, which controls the rudder.

Janissary A professional Muslim soldier.
Jib A triangular sail forward of the foremast.
Jolly Roger The pirate flag.
Junk A wooden vessel with rectangular sails used by pirates in the South China Sea.
Keel The bottom of a vessel's hull.
Keelhaul A sailors' punishment in which the accused is hauled on ropes under a vessel's keel.
Ketch A small sailing vessel or boat with two masts.
Lateen-rigged A vessel with triangular sails mounted on a yard hoisted to the top of a low mast.
Latitude Position north or south of the equator.
Lee On the side or direction away from the wind.
Letter of marque A licence authorizing a privateer to plunder enemy vessels.
Log book A written record of a vessel's progress.
Longboat The wooden single-masted vessel in

which the Vikings sailed.
Longitude Position east or west of the equator.
Mainsheet A rope at the bottom corner of a mainsail.
Man-of-war A large naval warship.
Marlinspike A pointed tool used to unravel ropes, sometimes used as a weapon, like a dagger.
Maroon To abandon a person on a remote island or other desolate place.
Mate A junior rank on board ship.
Mizzenmast The mast closest to the stern of a vessel.
Mutiny A rebellion among the crew of a vessel, often resulting in a takeover of command.
New World Sixteenth-century term for the Americas.
Piece-of-eight A Spanish silver coin worth eight *reales*.
Pink Eighteenth-century name for a *fluyt*, but varying in the number of masts.
Pirate One who seeks plunder on the high seas.
Plunder Treasure that is taken from a captured vessel or town, etc.
Port The left-hand side of a vessel.
Press gang A party of sailors who went ashore to force others to join their crew.
Privateer A seafarer who plundered enemy shipping under the authority of a letter of marque, or a ship used for such plundering.
Prize An item of plunder, especially a ship.
Ratlines The horizontal ropes attached to the shrouds to enable them to be climbed by a crew.
Rigging The ropes attached to the sails and masts of a vessel.
Schooner A small sailing vessel with two or three fore-and-aft rigged masts, sometimes also square topsails.
Scurvy A disease caused by lack of vitamin C.
Shallop A large vessel with one or more gaff-rigged masts.

Sheet A rope attached to the lower corner of a sail.
Ship A vessel with three or more square-rigged masts.
Shrouds The ropes supporting the mast or the topmast.
Sloop A small, single-masted sailing vessel with a fore-and-aft rigged mainsail and fore-and-aft rigged jib foresail (sometimes two- or three-masted vessels also).
Smuggler A person who illegally imports goods without paying duty (tax) on them.
Snow A two-masted square-rigged vessel with a spanker mounted aft of the mainmast.
Spanish Main Spain's possessions in the Americas, later applied to the islands and waters of the Caribbean.
Spanker A gaff-rigged sail.
Spar A length of wood forming a mast or yard.
Splice To weave two ropes together.
Square-rigged A vessel with square sails set on horizontal yards at right angles to the deck.
Starboard The right-hand side of a vessel.
Stern The rearmost part of a vessel.
Sterncastle A raised deck towards the rear of a vessel.
Swashbuckler A pirate of the dashing, romantic variety (a word introduced in the nineteenth century).
Sweep A long oar.
Swivel gun A small cannon mounted on a swivel.
Tack To change direction at sea by steering into the wind until the wind blows from the other side of the vessel.
Topmast The uppermost part of a mast.
Topsail A sail mounted on a topmast.
Waggoner A book of sea charts.
Weigh To haul up (an anchor).
Yard A horizontal length of wood to which the top of a sail is attached.
Yardarm One of the ends of a yard.

INDEX

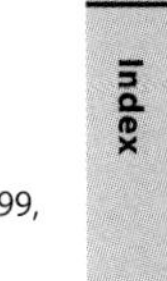

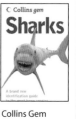

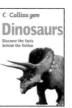

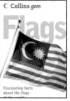

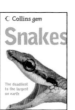